GRAHAM HOYLAND

MERLIN

THE POWER
BEHIND
THE SPITFIRE,
MOSQUITO
AND
LANCASTER

WILLIAM
COLLINS

William Collins
An imprint of HarperCollins*Publishers*
1 London Bridge Street
London SE1 9GF

WilliamCollinsBooks.com

HarperCollins*Publishers*
1st Floor, Watermarque Building, Ringsend Road
Dublin 4, Ireland

First published in Great Britain by William Collins in 2020
This William Collins paperback edition published in 2021

2022 2024 2023 2021
2 4 6 8 10 9 7 5 3 1

A catalogue record for this book is
available from the British Library

ISBN 978-0-00-835930-0

Typeset in Minion Pro
Printed and bound in Great Britain by
CPI Group (UK) Ltd, Croydon

MIX
Paper from
responsible sources
FSC™ C007454

This book is produced from independently certified FSC™ paper
to ensure responsible forest management.

For more information visit: www.harpercollins.co.uk/green

To the many, who designed, built and
serviced the Rolls-Royce Merlin engine

CONTENTS

Introduction   1

CHAPTER ONE   5
Non est ad astra mollis e terris via

CHAPTER TWO   21
Nobody ever understands what a pioneer is doing

CHAPTER THREE   29
Prometheus's gift

CHAPTER FOUR   37
Primum movens

CHAPTER FIVE   47
When Mr Rolls met Mr Royce

CHAPTER SIX   77
The Best Car in the World

CHAPTER SEVEN   89
Alis aquilae

CHAPTER EIGHT 113

A Rolls in the desert was above rubies

CHAPTER NINE 123

Racing improves the breed

CHAPTER TEN 133

Air racing's golden age

CHAPTER ELEVEN 157

The female of the species is more deadly than the male

CHAPTER TWELVE 177

Si vis pacem, para bellum

CHAPTER THIRTEEN 201

Guernica

CHAPTER FOURTEEN 211

The Spitfire is typically British. Temperate, a perfect compromise of all the qualities required of a fighter, ideally suited to its task of defence

CHAPTER FIFTEEN 239

It is perhaps as difficult to write a good life as to live one

CHAPTER SIXTEEN 271

The pity of war

CHAPTER SEVENTEEN   289

*Perfection is finally attained not when there is no
longer anything to add, but when there is no longer
anything to take away*

CHAPTER EIGHTEEN   301

*For they have sown the wind, and they shall reap
the whirlwind*

CHAPTER NINETEEN   319

Less is more

Epilogue   331
In Memoriam   333
Acknowledgements   335
Notes   337
Bibliography   347
Image Credits   351
Index   353

INTRODUCTION

'The Rolls-Royce Merlin was simply an astonishing engine. There have undoubtedly been aero engines that were better designed; it is possible but unlikely that there have been some that were better made; but beyond a shadow of a doubt there has never been another engine more thoroughly, continuously, aggressively, successfully and amazingly developed than the Merlin.'[1]

The story of the aero engine is as rich and strange and wonderful as anything found in the science and art of the Renaissance. In just 50 years, driven by two world wars and intense national competition, piston engines increased from a couple of horsepower to several thousands. In so doing they realised the eternal dream of humanity: the surreal experience of flight.

Far from the oily and inconvenient machines of popular imagination, these piston engines were dynamic sculptures of polished metal, designed using the intense application of physics, mathematics, metallurgy, chemistry and endless rigorous experimentation. The breakage of the tiniest part of many hundreds could lead to disaster and death. The failure of an engine design could lead to the loss of a war, because the Second World War above all others was decided through air power: all the decisive campaigns were won or lost by piston-engined fighters and bombers. The winner would be the side that could build the most powerful and reliable aero engines, and in this war the winner would take all.

The children ran around the school playground, arms extended, machine-gun thumbs out, shooting down their opponents in mock dogfights. The Second World War was only 20 years distant and most of their fathers had served in the forces. There was a wartime RAF station on the hill. I was reading the Biggles books by then and the Spitfire was the magic name on everyone's lips. The noises we were making, I now realise, were a childish imitation of the Merlin engine.

Later, thanks to help from kindly adults, I became interested in repairing piston engines. I marvelled at the finely carved pieces of metal that I knew grew into weights of over a ton at high speed. I saw the consequences of engines that burst from too many revs or seized due to the lack of oil.

Then one evening in the Peak District, many years later, I heard a distinctive drumming roar in the sky and, looking up, saw the elegant elliptical wings. It was a lone Spitfire heading home into the sunset after a day at a display somewhere in the South. I became curious. What was a Merlin engine? Why was it held in such esteem by aviation historians? Was it true that this was the engine that won the war? And had a wealthy woman really been behind both the engine and the aeroplane?

The Rolls-Royce Merlin was the aero engine that powered the Supermarine Spitfire, the Hawker Hurricane, the de Havilland Mosquito and the Avro Lancaster bomber, the aircraft that together turned the tide of the Second World War.

Eighty years ago, the distinctive roar of Derby-built Merlins was heard over the fields of southern England during the summer of 1940 as 'the Few' fought hordes of German aircraft during the Battle of Britain. Without such a powerful and reliable power unit at such a crucial time Britain would have lost the battle for the skies and the war could well have been won by the Axis powers.

It nearly didn't happen. At the last minute a wealthy benefactress, Lady Lucy Houston, had to provide the funds for the Rolls-Royce 'R' racing engine that sired the Merlin. Her money helped to nurture the new power unit. However, the Merlin had growing pains, being troublesome in its development and slow to come to maturity.

This story of powered flight is a story of our species at its best and worst: human inventiveness versus intellectual theft; human persistence versus jealousy and duplicity; human patriotism and courage versus hatred and dogma.

CHAPTER ONE

Non est ad astra mollis e terris via –

There is no easy way from the earth to the stars

As a child Albert Einstein was given a Russian toy steam train by a favourite uncle. Three years after winning the Nobel Prize for Physics he wrote to Caesar Koch to thank him once again. Einstein, the author of the theory of relativity, maintained that it was the steam engine that initiated his interest in science.

Steam engines are easy to understand and easy to love, as Thomas the Tank Engine shows. Fuel, either alcohol in the case of Einstein's toy or coal in a full-sized engine, is burned outside a water-filled boiler. This is external combustion. Steam is generated and tries to expand a thousandfold in volume. It passes through pipes and valves into a cylinder, where it pushes a piston down. The piston is connected by a rod to a crank which drives a wheel round, which drives the steam engine along. You can see the whole process going on, together with delightful smells of fuel, hot oil and smoke and the vision of an animate machine with long metal legs and pumping lungs.

The brilliance of the invention of the internal combustion piston engine was that it dispensed with a man shovelling coal,

The Langley 'Aerodrome' in flight – just.

the filthy firebox, the water, steam and heavy boiler, and instead burned petroleum fuel *inside* the cylinder. Everything else – pistons, connecting rods, crankshafts and flywheels – remained recognisably related to steam engines. Sadly though, these engines are hard to appreciate in the same way as steam. If you lift the engine cover of a car all you can see is an immobile cylinder block made of aluminium or cast iron, and all you can hear is the whirring and clicking of innumerable parts: valves, camshafts, pistons and crankshafts. It's all going on inside. Drive or fly behind a piston engine, though, and listen to the operatic howl of many cylinders, and your soul may begin to be stirred by something beyond words.

The story of the piston aero engine is surely the most romantic story in engineering history. Heavier-than-air machines were dreamed of but literally could not take off until light and powerful internal combustion engines were built. Steam engines were simply too heavy.

Mankind had always dreamed of flight. And made up stories about flight. According to the scholar Ben Sherira, flying carpets were issued to readers in the ancient library of Alexandria (*c.* 283 BCE) in exchange for their slippers. Reclined on their carpet, students were able to reach the highest shelves and hovered near the ceiling, engrossed in their studies. Later, in the *Baopuzi* text (320 CE), the Chinese master Ge Hong described the principles of ascending into the vast inane: 'some have made flying cars with wood from the inner part of a jujube tree, using ox or leather straps fastened to blades so as to set the machine in motion.'[1] He seems to be describing a Jin dynasty helicopter.

In around 559 the Emperor Wenxuan of Northern Qi decided to conduct experiments with flight. Volunteers failed to come forward, so prisoners were forced to leap off a tower attached to man-carrying kites. There was only one survivor, Yuan Huangtou of Ye, who successfully managed to glide over the city walls and land safely. He was later executed.[2] In the eleventh century Eilmer of Malmesbury, an English Benedictine monk, attempted to emulate Daedalus of the Greek myth by attaching wings and leaping off a hill near the abbey. He flew for a furlong (200 metres), but crashed more like Icarus, breaking both legs, which rendered him lame for life. 'I lacked a tail,' he remarked ruefully later, showing that at least he had a grasp of aeronautics.[3]

These stories make Leonardo da Vinci's interest in flight seem rather late in the day. In 1485 he drew a man-powered rotor that could not have flown, as the body of the machine would have rotated in the opposite direction to the rotor. He drew other man-powered flying machines with flapping wings which are unlikely to have got very far. However, he also designed a hang glider which could have flown, and indeed flying examples have been built. But what Leonardo needed was more power, and he realised this. He didn't know that what he *really* needed was a petrol engine, and he would probably have swapped a couple of dozen sketches for a really good lawnmower.

Balloons were the first successful aircraft, and they relied on weighing less than the air they displaced, but they were large, delicate and slow, blown across the sky like tender elephants. Balloonists wanted to progress in their own choice of direction instead of that of the wind, so various methods of propulsion were tried. Human-powered flapping wings were employed in 1784 by the balloonist Jean-Pierre Blanchard who, helped by the wind, managed to cross the English Channel, flapping manfully and steering with a bird-like tail. In 1852 another Frenchman, Henri Giffard, combined a highly explosive hydrogen balloon with a disturbingly flammable steam engine, which proved able to perform limited manoeuvres but was not powerful enough to fly against the wind. This was clearly not the way to go.

Countless attempts were made to build a heavier-than-air machine, but the mistake the inventors usually made was to try to copy the flapping motion of birds' wings. One honourable exception was the Frenchman Alphonse Pénaud. He pioneered the use of rubber-band-powered model aircraft with fixed wings and a propeller rather like the children's toys available today. His model aeroplane of 1871, which he called the Planophore, was the first aerodynamically stable flying model. Poor Pénaud tried to attract interest and funds for a full-sized machine, but failed. He committed suicide in despair, aged just 30.

The immortal Robert Hooke, inventor of the sash window and theorist of springs, realised that manpower was insufficient and built a spring-powered winged ornithopter. The flapping wings theme was pursued by the inventive Frenchman Gustave Trouvé, whose model of 1890 flew for 80 metres in a demonstration for the French Academy of Sciences. His wings were flapped by gunpowder in cartridges exploding in sequence. We can only be grateful that his style of flight did not catch on. Someone had to apply some proper science.

At last someone did exactly that: the Yorkshireman Sir George Cayley (1773–1857).[4] He observed the soaring flight of seagulls and realised that it was the angle and shape of the wings that produced lift. Flapping produced propulsion, not lift. This was a moment of epiphany, similar to the inspirational moment that produced the wheel axle, the screw-thread or the lever.

Cayley then described four paired forces acting upon a flying machine: weight and lift, thrust and drag. He had a silver disc engraved with a design for an aircraft, with the four forces engraved on the reverse side of it. He decided that fixed-wing flight was easier to achieve than trying to imitate the flapping wings of birds, and that cambered wings set at an angle to the wind would provide lift. He realised the importance of the dihedral angle: the upwards tilt of aircraft wings that provides stability. He also predicted that sustained flight would not be achieved until a lightweight source of power could be invented to provide thrust and lift. And he even attempted to invent a suitable internal combustion engine.

Cayley didn't know that in 1806 two French brothers, Nicéphore and Claude Niépce, had made what was probably the world's first internal combustion engine, which they called the Pyréolophore. Bizarrely, it was fuelled by lycopodium powder (dried spores of the lycopodium clubmoss plant) and coal dust. It worked rather like a toy steam pop-pop boat but would only operate in water. The brothers' patent was signed by Emperor Napoleon Bonaparte, but the brothers failed to capitalise on their engine. Claude moved to Kew, London, to spend all the family fortune on trying to sell the Pyréolophore and descended into delirium, and his brother got on with inventing photography.

Cayley wrote a treatise entitled *On Aerial Navigation* and continued his experiments. In the pursuit of lightness for his aircraft he reinvented the wheel: wire-spoked wheels, which suspend the load from wires in tension rather than using heavy wooden spokes in compression. His wheels are still in use today on bicycles. He realised that he lacked a suitable source of power and tried to build internal combustion engines running on gunpowder or hot air, but the technology still eluded him.

Finally, at the age of 75, Sir George and his helpers at Wydale Hall flew a full-sized glider across Brompton Dale in 1853. According to one author the pilot was his ten-year-old grandson George. So a boy could well have been the first person to be carried by a modern fixed-wing heavier-than-air flying machine.

Sir George Cayley was one of the most important pioneers of flight, and was the first to take a truly scientific approach to the study of aeronautics. He really deserves his title of 'father of aeronautics'. If he'd only had a suitable lightweight engine the world's first pilot-controlled powered flight could have taken place over the moors of North Yorkshire in the middle of the nineteenth century.

Then in 1876, after the efforts of dozens of engineers and scientists, a viable internal combustion piston engine finally arrived. This engine had been developed by a Belgian, Jean Joseph Étienne Lenoir, but his machine was heavy and inefficient, as it followed the old steam engines in its design. Running on lighting gas, it produced only 2 horsepower for 18 litres of cylinder capacity. This design was then considerably improved by the German Nikolaus Otto, who perfected a way of compressing air and fuel and igniting it to provide power using four distinct strokes of the piston. This was the inventive step that made powered flight possible.

The way an Otto-cycle four-stroke petrol engine works is the same in a Merlin or a modern car engine (my apologies to those readers who already know this). A piston, which looks like a soup can with one end removed, slides up and down a tube or cylinder. The piston is connected to a rod called, unsurprisingly, a connecting rod. This is attached to a crankshaft so that the piston sliding up and down pushes the crankshaft around rather as a cyclist's leg pushes a pedal crank around. The crankshaft can be connected to car wheels or to an airscrew propeller.

Above the piston is where the magic of internal combustion takes place. The top of the cylinder is closed but is provided with an inlet 'poppet' valve to let in a mixture of air and fuel as the piston descends on the first, or inlet, stroke. This valve then closes when the piston reaches the bottom and starts up again on the second, or compression stroke. The gaseous mixture of air and fuel is now squeezed tightly and is highly explosive. At the top of the compression stroke a sparking plug ignites the mixture and it duly explodes. The piston is now pushed down on its third stroke, the power stroke. On its way back up an exhaust valve opens, and the hot gas is forced out by the ascending piston on the fourth, exhaust, stroke.* Just think 'suck, squeeze, bang, blow'.

Otto's inventive step was to manage these four distinct phases inside a cylinder with a piston travelling up and down, and he did this by arranging the valves to open at specific times during two revolutions of the crankshaft, or four strokes of the piston. This was done by using a camshaft, which looks something like a knobbly stick, to push the valves open with oval-shaped lobes. By making the camshaft rotate only once for every two rotations of the crankshaft the valves open at the correct times.

Now speed the whole engine up to make more power. A modern 1-litre car engine can run its crankshaft at 6,000 revolutions per minute (rpm), which means that it rotates 100 times in a second. So a piston is going down, stopping, reversing direction and going up again 200 times a second, and the four Otto cycles are happening in milliseconds. A typical car piston weighs around 0.3 kilograms at rest, but when changing direction at 6,000 rpm it would effectively 'weigh' 900 kilograms! That's as much as a car. No wonder pistons are made of immensely strong and light aluminium alloy. A 27-litre Rolls-Royce Merlin revolves at only 3,000 rpm, but even then the load on the crankshaft main bearings is around nine tonnes – the weight of a bus.

---

* A glance at an online animated graphic will explain it all: https://en.wikipedia.org/wiki/Internal_combustion_engine.

Virtually all petrol and diesel engines share the same kinds of pistons, cylinders and valves. There may be as many as 36 cylinders in various configurations, and the Merlin and most modern cars have two inlet valves and two exhaust valves per cylinder. A typical car engine will have four cylinders in a row above the crankshaft, but a rotary aero engine has the cylinders and propeller whirling around a stationary crankshaft. A *radial* aero engine has stationary cylinders arranged in a star shape, with a rotating crankshaft driving the propeller. And the Rolls-Royce Merlin had 12 cylinders arranged in two rows of six sharing one crankshaft: a V12 configuration. There are advantages and disadvantages for all these formats.

A word about horsepower. This is a much-abused measure of the power of an engine, originally invented by the Scottish engineer James Watt in the late eighteenth century to compare the power of mine horses with his steam engines. So does one horse produce one horsepower? Well, no. The maximum power of a horse has been measured at 14.9 horsepower for a few seconds, and as for humans, the Jamaican sprinter Usain Bolt produced 3.5 horsepower during his 100-metre world record in 2009. How so?

The point is that you would need to keep around 15 horses to provide a continual 1 horsepower: after all, the poor creatures need to be rested and fed. Watt's estimate was based on mine horses working a four-hour shift.

A story goes that the measurement was created by Watt when one of his customers, a rascally brewer, demanded a steam engine that could match his horse. He chose the strongest horse he could find and drove it to exhaustion. Watt, well aware of what was going on, accepted the brewer's challenge and built an engine that exceeded the figure achieved by the poor horse. When calculating the measurement Watt allowed for the fact that a single horse cannot work day and night and adjusted accordingly.

The accepted measure of one mechanical horsepower is the ability to lift 550 pounds one foot in one second, or 745.7 Watts.* You can calculate horsepower by multiplying the torque (or twisting force) by revolutions (the distance travelled) and then dividing by 5,252 (the time – don't ask).

At least the suffering of countless draught horses was alleviated by engines. But horsepower as a measure has attracted so many qualifications and fudges that it has to be used with caution.

Nikolaus Otto and his factory manager Gottlieb Daimler disagreed about what to do next with his four-stroke invention. Otto wasn't really interested in transport and wanted his engine to replace large stationary steam engines. Daimler wanted to make small engines suitable for bicycles or carriages. As a result, Daimler left Otto and set about evading his patents so that he could avoid paying royalties. By a subterfuge involving the 'previous art' of a patent granted in 1862, Daimler overturned Otto's patent and went on to build an empire. In 1885 he and Wilhelm Maybach built a small single-cylindered half-horsepower engine and put it in a bicycle. So the first person in the world to ride an Otto-cycle motorised vehicle was Adolf, Daimler's 14-year-old son.

We now have to uncover an extraordinary attempt to unfairly claim the first powered flight by that most upright and sober of American institutions: the Smithsonian. Samuel Langley was the founder of the Smithsonian Astrophysical Observatory, but he then took up an interest in aviation, building rubber-band-pow-

---

* The Watt is an SI unit created in 1882 to measure power – usually electricity. It was named for James Watt.

ered model aircraft based on Pénaud's designs. Colleagues noticed that his models did not seem to stay in the air very long. Then he managed to make steam-engined flying models. Around this time Rudyard Kipling was visiting his friend (and soon to be president) Teddy Roosevelt in Washington. He encountered the professor with the passion for toy aircraft:

> Through Roosevelt I met Professor Langley of the Smithsonian, an old man who had designed a model aeroplane driven – for petrol had not yet arrived – by a miniature flash-boiler engine, a marvel of delicate craftsmanship. It flew on trial over two hundred yards, and drowned itself in the waters of the Potomac, which was cause of great mirth and humour to the Press of his country. Langley took it coolly enough and said to me that, though he would never live till then, I should see the aeroplane established.[5]

This mockery by the press may have provoked the Smithsonian into their later claims of success on the part of their colleague. Langley's models flew further and further until his No. 6 flew for nearly a mile. The Smithsonian then granted him $20,000 to develop a full-sized man-carrying machine. To this the US War Department added $50,000, a huge sum of money at the time. The stakes were correspondingly high.

Langley called his pilot-carrying machine 'The Aerodrome', which translates as 'air runner' in Greek. The name has lived on to denote the runways that aircraft fly from. The younger man Langley hired to be the pilot was a brilliant American engineer, Charles Manly. And realising that the power-to-weight ratio of a steam engine was too low to propel a full-sized aircraft, Langley looked for someone to build him one of the exciting new Otto-cycle engines.

Professor Langley commissioned a Hungarian-born engineer, Stephen Balzer, to build him an engine for the Aerodrome. Balzer had built his first motor car in New York, a car that featured a lightweight rotary three-cylinder engine which he thought he could enlarge for the Aerodrome. Car and motorcycle engines are usually too small for aircraft. His five-cylindered prototype was disappointing, only producing 8 to10 horsepower compared with the minimum of 12 hp that Langley needed. It only ran for a few minutes. As many engineers have found to their cost, simply enlarging a design doesn't always work.

Langley asked his engineer/pilot Charles Manly to intervene, and so he took on the mammoth job of redesigning the engine. A trip to Europe in 1900 to speak to other engine designers convinced him that the rotary design was not the way to go. So he designed the engine with *fixed* radial engine cylinders. And he designed a masterpiece.

Charles Manly made, largely with his own hands, a five-cylindered four-stroke radial engine, nearly 9 litres in total cylinder capacity. Five cylinders would give smooth running, and they were laid out somewhat like Leonardo's Vitruvian Man, in a star shape.

He put a cast-iron liner into each spun-steel cylinder, something his local machine shops said couldn't be done, and so he did it himself. He also welded steel water jackets around the cylinders for cooling and brazed ports on for the valves. In so doing he permanently damaged his eyesight.

Why the need for cooling? It may shock the reader to be told that the average petrol engine is as little as 25 per cent efficient. That means that up to 75 per cent of the expensive fuel we buy disappears as wasted heat from the cylinders and down the exhaust. This has enormous consequences now in a warming world, but the problem for an early engine designer was how to remove all this excess heat without melting pistons and cylinders together in a red-hot mass. One solution was to put cooling fins on the outside of the cylinders, and this worked well enough for small motorcycle engines: we call this air-cooled. But when there are rows of cylinders, the rearmost do not receive enough cooling draught. The other solution is to place a jacket of cooling water around the cylinder and pump the near-boiling water away to a radiator well away from the engine: water-cooled. This was the route that Manly took.

The rest of his engine was no less remarkable. Manly had to make every part himself, as there were no lightweight aircraft-engine accessories available off the shelf. His ignition system using a high-voltage coil and spark distributor was the first of its kind. His sparking plugs featured a platinum electrode, a choice of material years ahead of time. But his carburettor was a chamber filled with wooden balls soaked in petrol which produced a vapour that was drawn into the cylinders. This was like someone building a Concorde out of corrugated iron: Manly simply didn't have specialised materials such as aluminium, and so he had to make the crankcase out of steel and the pistons out of cast iron. He invented a drum-type cam to open the valves and evolved the idea of having a master connecting rod and the remaining rods as slaves. The results were outstanding: the performance of this engine was prodigious. Aircraft builders are obsessed with the weight-to-power ratio of the engines they buy for obvious reasons: every extra pound or kilogram in the engine has to be supported by thin air. For a dry weight of 125 pounds, Manly's engine gave over 52 horsepower, a weight-to-power ratio of 2.4 pounds per horsepower, which was not beaten until 1916 by an engine designed and built in a fully equipped factory during wartime. Furthermore, his engine ran reliably for ten hours without getting hot and bothered.

Not only did Manly produce the world's first purpose-made aircraft engine, he did it largely on his own. L. J. K. Setright,* the sage of aero engineering, wrote this: 'As an example of brilliant originality in design and virtuoso ability in workmanship, the Manly radial remains to this day one of the most outstanding aero-engines in history.'[6]

The sequel to this story is a sad one. After this jewel of an engine was fitted into Professor Langley's Aerodrome, the aircraft was loaded onto a catapult mounted onto a houseboat moored on the Potomac River. It was 7 October 1903. But the aircraft had been scaled up from models and was structurally far too weak. Control was virtually non-existent, with two sets of wings in tandem and a central rudder. Manly took the controls, the engine roared, the whole contraption slid down the ramp … and promptly flopped into the water. Manly had to be rescued.

On the second attempt on 8 December Manly tried again, and this time the Aerodrome nearly killed him. It collapsed into the water again, trapping him under a tangle of wires and canvas. He was dragged out just in time.

Professor Langley made no further attempts, and the whole expensive project was ridiculed by the press. Then just nine days afterwards on 17 December 1903, Wilbur and Orville Wright conducted four successful flights near Kitty Hawk, North Carolina with an engine of their own construction.

* * *

* Setright was the paragon of technical writers: car journalism's Prospero to Jeremy Clarkson's Caliban. Concert musician, Jewish scholar and epicure, in appearance he was a gaunt Old Testament prophet draped in a Savile Row suit. More English than the English, he smoked Black Russian Sobranie cigarettes and drove a Bristol motor car.

We are not quite finished with the Aerodrome. Professor Samuel Langley was clearly an honourable man, and when a colleague was found to have been embezzling funds from the Smithsonian he held himself responsible and refused his salary. The pressure proved too much, he suffered a stroke and died in 1906.

In a dubious alliance, the Smithsonian allowed the aircraft manufacturer Glenn Curtiss to make extensive modifications to the Aerodrome to enable it to fly. Curtiss wanted to prove it was the first flying machine so that he could defeat the Wright brothers' patent lawsuits against him after he appropriated their ideas. And the Smithsonian wished to salvage their deceased secretary's aeronautical reputation. After a few short hops by Curtiss during which daylight was seen under the Aerodrome, the Institution was sufficiently emboldened to display the machine in its museum. 'The first man-carrying aeroplane in the history of the world capable of sustained free flight,' the placard boasted. 'Invented, built and tested by Samuel Pierpont Langley in 1903.' This relegated the Wright Brothers' 1903 Kitty Hawk Flyer to also-ran status.

'It was a lie pure and simple, but it bore the imprimatur of the venerable Smithsonian and over the years would find its way into magazines, history books, and encyclopaedias, much to the annoyance of those familiar with the facts,' wrote the aviation historian Fred Howard.[7] 'To Orville Wright it was more than an annoyance. It was the culmination of Glenn Curtiss's campaign to demean and devalue all that the Wright brothers had accomplished.'

Orville Wright was quite rightly furious (his brother Wilbur had died two years previously), and he accused the Institution of misrepresenting American aviation history. He found out the extent of the Curtiss modifications from a close friend of the Wright brothers, Griffith Brewer, who had photographed the tests. He then refused to donate the original Flyer to the Smithsonian, instead giving it to the Science Museum of London, which had a more objective view of aviation history. The feud eventually ended, but not until the Smithsonian published details of the Curtiss modifications to their Aerodrome and recanted their claims for precedence.

At last, the executors of Orville's estate signed an agreement for the Smithsonian to purchase the Wright Kitty Hawk Flyer for one dollar. At the insistence of the executors, their agreement included conditions for display of the aeroplane: if any associated institution dared to claim precedence for any other aircraft, then the Flyer would be forfeited to the heir of the Wright brothers.

CHAPTER TWO

*Nobody ever understands what
a pioneer is doing*[1]

So how did two humble bicycle mechanics from Dayton, Ohio change history? A host of French, British and German engineers and inventors knew that powered heavier-than-air flight was coming and were racing to be the first. There was also a horde of frauds, stuntmen and profiteers. They all thought that the new petrol engine was going to provide the breakthrough they craved, but what they didn't know was how they were going to control the aircraft once it left the ground. And that is exactly what Orville and Wilbur Wright figured out.

Like Einstein's steam engine, it's another story of an inspirational toy: when they were boys their father Milton brought home a toy flying machine. It was a form of helicopter based, like Professor Langley's, on a design by Alphonse Pénaud. It was powered by a rubber band that drove a rotor. Wilbur and Orville were fascinated and played with it until it broke, and so they built their own. Like Albert Einstein, they later maintained that it was this toy that initiated their lifelong interests.

'Surely we're flying this in the wrong direction?' – Wilbur Wright on the Wright Flyer in 1903.

After a career together building a printing press and running a newspaper, the brothers, neither of whom married, turned their attention to the new craze of bicycling and built up their own safety bicycle manufacturing company. This funded their new interest in flight, triggered by Professor Langley's steam-driven model aircraft flights at the Smithsonian and the exploits of the German hang-gliding pilot Otto Lilienthal, who was killed in August 1896. Wilbur said of him: 'Lilienthal was without question the greatest of the precursors, and the world owes to him a great debt.'

The brothers decided that pilot control was vital and first learned to build gliders. They evolved the idea of banking, or leaning the aircraft into turns – rather like riding a bicycle. They achieved this by twisting the ends of the wings, or 'warping' as they called it. They experimented with gliders at an area of sand dunes called Kill Devil Hills at Kitty Hawk, North Carolina, using a local boy as pilot, as did Sir George Cayley. They then tested 200 types of wing in their home-made wind tunnel, and also figured out the best kind of propeller, which they eventually realised was a form of rotating wing.

After their experience with bicycles and studying the observations of predecessors such as Sir George Cayley, the Wright brothers realised that pilot control was essential for such a dangerous and unstable machine as an aircraft. Wilbur Wright acknowledged Cayley's importance to the development of aviation: 'About 100 years ago, an Englishman, Sir George Cayley, carried the science of flight to a point which it had never reached before and which it scarcely reached again during the last century.'[2] The Wrights' inventive breakthrough was three-axis control, which is still used to this day. In their earlier unpowered gliders they learned how to control roll, pitch and yaw, partly by twisting the wings, partly by using a rudder.

Compared with Manly's jewel their engine was fairly unremarkable, but it did the job. The first engine to propel a heavier-than-air machine was not the best of those early engines, but it *was* the first. They could not buy one that would do the job, so they set about building one out of local materials. Largely following the contemporary 1900s automobile practice of '4x4x4', it had four cylinders measuring 4 inches in diameter with a 4-inch piston stroke. These were laid flat in a row and a four-throw crankshaft was carved out of a solid billet of steel by the Wrights' machinist, Charlie Taylor. Here's Taylor's story in his own words:

> We didn't make any drawings. One of us would sketch out the part we were talking about on a piece of scratch paper, and I'd spike the sketch over my bench. It took me six weeks to make that engine. The only metal-working machines we had were a lathe and a drill press, run by belts from the stationary gas engine.

This single-cylindered engine was one they had built earlier to test their skills:

> The crankshaft was made out of a block of machine steel 6 by 31 inches and 1-5/8 inch thick. I traced the outline on the slab, then drilled through with the drill press until I could knock out the surplus pieces with a hammer and chisel. Then I put it in the lathe and turned it down to size and smoothness …

This crankshaft was literally hacked out from the solid by Taylor. It lacked any balance weights, so the vibration must have been numbing for the pilot, lying prone next to the engine.

… the ignition was the make-and-break type. No spark plugs. The spark was made by the opening and closing of two contact points inside the combustion chamber. These were operated by shafts and cams geared to the main camshaft. The ignition switch was an ordinary single-throw knife switch we bought at the hardware store. Dry batteries were used for starting the engine, and then we switched onto a magneto bought from the Dayton Electric Company.[3]

The engine's capacity was 3.3 litres. It had an overhead camshaft, which opened the exhaust valves only, as the spring-loaded inlet valves were sucked open automatically by the descending pistons, which was common practice at the time.

What was remarkable was the material used for the crankcase. Unlike Manly, the Wrights managed to find the latest precipitate-hardened aluminium, and this was also a first in aircraft construction. They contracted a local Dayton foundry, the Buckeye Iron and Brass Works, to cast the aluminium crankcase. Buckeye acquired their raw aluminium from the nearby Pittsburgh Reduction Company, which was renamed Alcoa in 1907 and later became the world's leading producer of aluminium. Here's Charlie Taylor again:

The body of the first engine was of cast aluminum and was bored out on the lathe for independent cylinders. The pistons were cast iron, and these were turned down and grooved for piston rings …[4]

Amazingly, the engine had no fuel pump and no carburettor, but a crude form of fuel injection which allowed a trickle of petrol to leak into a water-heated inlet manifold where it vaporised and was sucked in through the automatic inlet valves. So the pilot had no throttle to control the engine power, just an ordinary tap, and therefore the engine ran at full speed or not at all. Presumably the pilot was otherwise occupied in learning how to fly the world's first aeroplane on its very first flight.

All of this hard work and inventiveness resulted in a dry weight of 179 pounds, quite a bit more than Manly's engine at 125 pounds. The specific power was 15 lb/hp instead of his 2.4 lb/hp. It also only produced 12 horsepower instead of Manly's 52 horsepower, and it couldn't produce that power for long. After a few minutes heat would build up and the inlet manifold would get too hot, reducing the density of the air and thus reducing power to around 8 horsepower.[5] It was enough power – but only just.

The Wright engine was first run on the bench on 12 February 1903, but on the very next day (Friday the 13th, as it happened) it overheated and seized up on a test run. Overheating was to be a constant bugbear of aero-engine manufacture over the next century, as we will discover. New castings had to be ordered from the foundry, and Charlie had their engine rebuilt and ready to go in early June.

What was even more impressive than building an engine literally from scratch is that the brothers realised the importance of running the two propellers slower than the engine and revolving them in opposite directions to counteract gyroscopic forces (these two facts didn't dawn on some aircraft manufacturers until well into the Second World War). They did this by the simple expedient of using long bicycle-style chains and different-sized sprockets to gear down the 8-foot (2.5-metre) spruce propellers, then simply twisting just one of these chains into a figure-of-eight to reverse the direction! The brothers' bicycle experience had certainly paid off. The result of all this was just 90 pounds (40 kg) of thrust (the most recent Rolls-Royce Trent jet engine generates 97,000 pounds (44,000 kg), well over a thousand times as much as that first aircraft engine).

This historic engine came to a sad end. After powering the Flyer on four flights at Kitty Hawk on 17 December 1903, it suffered serious damage, splitting the crankcase when a gust of wind overturned the aircraft. It wasn't perhaps the best engine built for pioneer flight – that was Manly's. But it was the first engine to fly.

We are nearly at the point of lift-off. In 1903 the brothers built their aeroplane, the Wright Flyer, adding Charlie Taylor's engine. The whole aircraft cost less than a thousand dollars, in contrast to the $70,000 total of government and Smithsonian funds spent by Professor Langley on his Aerodrome.

Finally, on 14 December 1903 Wilbur and Orville tossed a coin to see who would go first. Wilbur won the toss and the two brothers shook hands. John T. Daniels, a lifesaver crewman, was struck by the feelings between the two brothers: 'After a while they shook hands, and we couldn't help notice how they held on to each other's hand, sort o' like they hated to let go; like two folks parting who weren't sure they'd ever see each other again.'[6]

The two brothers wouldn't see each other again, not in a flightless world. A few minutes later Wilbur lay down on the lower wing, the Flyer's engine roared, the aircraft trundled down the launch rail and an aircraft took off under its own power for the first time in history. The Age of Aviation had begun.

CHAPTER THREE

Prometheus's gift

After a three-and-a-half-second flight the Wright Flyer lurched up steeply, stalled and crashed nose-first into the sands of Kill Devil Hills. Wilbur was slightly shaken, but the canard elevator at the front had taken most of the impact. He wrote home: 'the power is ample, and but for a trifling error due to lack of experience with this machine and this method of starting, the machine would undoubtedly have flown beautifully ... there is now no question of final success.'

So was this the first successful flight? Or not? This is a problem for historians who try to hammer solid markers into the ever-shifting sands of time. Geologists also like to mark the transition between geological eras with what they call the 'golden spike', referring to the ceremonial gold final spike that was used to join two railway tracks when they met in the middle of the US in 1869, forming the transcontinental railroad.

The two men made a far more successful flight on their next outing: on their next test day, 17 December 1903, the Wrights arrived on site in a freezing wind of 27 mph (43 km/h), and with

The first successful flight of the Wright Flyer, 1903. Wilber has just let go of the wing.

the damaged elevator repaired Orville took off at 10.35 a.m. The over-sensitive front elevator caused the Flyer to swoop up and down in a sickening way, but it landed safely after a flight of 120 feet (36 metres) in 12 seconds. This flight was the subject of the famous photograph (his first) taken by John T. Daniels. This is generally accepted as the first heavier-than-air flight in history. Significantly the aircraft landed at the same level as the take-off: it hadn't just floated off a hill like a glider.

Wilbur managed 175 feet (52 metres) on the second flight, again struggling for control. On the third flight Orville achieved 200 feet (60 metres) in 15 seconds. And the fourth and final flight of Flyer 1 was more controlled. Orville recorded what happened:

Wilbur started the fourth and last flight at just about 12 o'clock. The first few hundred feet were up and down, as before, but by the time three hundred feet had been covered, the machine was under much better control. The course for the next four or five hundred feet had but little undulation. However, when out about eight hundred feet the machine began pitching again, and, in one of its darts downward, struck the ground. The distance over the ground was measured to be 852 feet; the time of the flight was 59 seconds. The frame supporting the front rudder was badly broken, but the main part of the machine was not injured at all. We estimated that the machine could be put in condition for flight again in about a day or two.[1]

This was more like it. Wilbur Wright had managed nearly a minute of sustained, controlled flight. The engine had run reliably throughout. They had a photograph and five witnesses, and Flyer had done her duty. Unfortunately the fragile aircraft was then caught by a huge gust of wind and rolled over several times, despite desperate attempts to hang on to it. As we have seen, the engine crankcase was split in half, and that was the end of the engine. This is the aircraft that Orville donated to the Science Museum in London and that now resides in the Smithsonian.

The brothers now had to capitalise on their invention, and that proved much harder. The Wrights began applying for a patent, but it described the 1902 warping wings on their glider, not the 1903 powered Flyer. It took over three years to be granted, and this patent was to be the basis for the many patent-infringement suits which exhausted the brothers (and probably led to Wilbur's premature death at only 45). In 1904 they built another Flyer with an enlarged version of the same engine and managed to fly in a large circle for one and a half minutes, but still the aircraft was difficult to control. There was a reason for that.

An amateur might be forgiven for thinking that the Wright Flyer 1 was flying backwards.* It had no tail, but instead a large appendage at the front. The propellers were at the back, pushing forwards. The Wrights were trying to fly their contraption in the wrong direction, in fact the whole thing was dangerously unstable, as modern analysis shows. An attempt to re-stage the hundredth anniversary of the first flight on 17 December 2003, using an exact replica, failed because the pilot simply could not control it. It was like a flying shopping trolley.

The Wrights struggled with their reputation at first. They refused to show anyone their machine as they feared their ideas would be stolen – and in that they were right. In Europe, aviation enthusiasts were certain that the French were on the brink of powered flight.

* The Soviet Antonov A-2 biplane really could fly backwards. Its stall speed was so low that a pilot could hold it into a head wind near the ground at around 25 mph (40 km/h) and if the wind was strong enough, this manoeuvre would make the plane move backwards very slowly while still under full control. Biplanes were flown by the all-women crews of the Soviet 588th Night Bomber Regiment.

The pioneer French aviator Captain Ferdinand Ferber had even written to his fellow countryman Ernest Archdeacon, founder of the Aéro-Club de France: 'Do not let the aeroplane be achieved in America first!'

It seems hard to believe in this world of 24/7 online news, but the only press coverage the Wrights received in 1904 was an article written by one Amos Root in *Gleanings*, a bee-keeping magazine, alongside articles such as 'How to get bees into the cellar' and 'Mid-winter flights of cellared bees'. In 1905 the Wrights' local newspaper, the *Dayton Daily News*, finally reported Wilbur's 5 October flight of 24.5 miles – a world-shattering achievement – on page 9, alongside farming news.

The mainstream press refused to take the brothers seriously, and the *New York Herald* was positively insulting: 'The Wrights have flown or they have not flown. They possess a machine or they do not possess one. They are in fact either fliers or liars. It is difficult to fly. It is easy to say, "We have flown."'[2]

The Wrights tried to interest the US Army in their invention, and made demonstration flights. On 17 September 1908 a propeller disintegrated in mid-air and Orville suffered a dreadful crash. His passenger, Army observer Lieutenant Thomas Selfridge, was killed, and thus entered history as the first person to die in an aviation accident. Orville was badly injured.

Prophets are not without honour, except in their own country. It wasn't until Wilbur took Flyer Model A to Europe in May 1908 that the international aeronautical community at last believed the Wrights' claims. His first flight near Le Mans only lasted 1 minute 45 seconds, but he made banking turns and described a complete circle. During later flights he made figures-of-eight and performed manoeuvres surpassing anything rival pioneering aircraft could achieve.

The plaudits flooded in, at last. The sceptics apologised and praised the brothers. The public loved Wilbur's flight demonstrations, and when Orville and their sister Katharine joined him in 1909, they were briefly world-famous. Kings came to see the flights. Previous sceptics now hastened to make amends. The *Aérophile* magazine, which had once sneered, now hastened to proclaim that the flights 'have completely dissipated all doubts. Not one of the former detractors of the Wrights dare question, today, the previous experiments of the men who were truly the first to fly.'[3] And Ernest Archdeacon, who had once written that the French would make the first public demonstration of powered flight, now wrote: 'For a long time, the Wright brothers have been accused in Europe of bluff ... They are today hallowed in France, and I feel an intense pleasure ... to make amends.'[4]

All very gratifying, but the Wrights struggled to make a success of their business. The crashes began to mount up. All six Wright C model aircraft bought by the Army crashed, and by 1913 the death toll had reached 11. Eventually the military refused to buy aircraft with 'pusher' type propellers with the engine behind the pilot as it was apt to crush him when the aircraft crashed into the ground. The competition in Europe and the USA was ferocious, and in the end patent struggles proved too much for the fledgling business.

Many years later Orville struck a sad note about the horrors brought about by the bombers of the Second World War:

We dared to hope we had invented something that would bring lasting peace to the earth. But we were wrong ... No, I don't have any regrets about my part in the invention of the airplane, though no one could deplore more than I do the destruction it has caused. I feel about the airplane much the same as I do in regard to fire. That is, I regret all the terrible damage caused by fire, but I think it is good for the human race that someone discovered how to start fires and that we have learned how to put fire to thousands of important uses.[5]

Prometheus brought fire to humanity. It was up to humanity to know what to do with it.

CHAPTER FOUR

Primum movens –

'Prime mover' – a new engine for a new century

Those Magnificent Men in their Flying Machines was a 1965 comedy caper film that featured authentic flying replicas of early aeroplanes. In it, the English press baron Lord Rawnsley puts up a £10,000 prize for the winner of a 1910 London-to-Paris air race. Many of the pioneer aircraft and their piston engines can be seen flying, but the authenticity of the cars, aircraft and sets is somewhat let down by feeble performances and a predictable script. Watching the stunt pilots getting these contraptions into the air gives an idea of just how dangerous this new sport was. Landing them again was potentially lethal.

As the Wright brothers built up their flying hours they ran into an engine problem. As they rotate, in-line four-cylindered engines' pistons stop and start all at the same time, and as a result such an engine can never have perfect balance. The vibrations of their engine were so bad that they threatened the structure of the aircraft. More power was needed, too, so the

A nine-cylinder Gnome rotary engine, *circa* 1916. The inlet valves are hiding in the piston crowns.

brothers increased the power of the four-cylinder engine to 25 horsepower by 1906, and in 1908 they made a six-cylindered engine of 39 hp.

Six-cylindered engines are smoother because the crank throws are at 120° to each other instead of the 180° of a four-cylinder. This means that paired pistons are at equal distance from the centre of the engine (nos 1 and 6 cylinders, nos 2 and 5, nos 3 and 4), and they are always moving together, which results in good balance in the reciprocating mass. The overlapping torque is generated every 120°, which helps for smoothness, too. However, you cannot keep adding cylinders in a line indefinitely – the crankshaft grows so long that it starts to twist and vibrate. This is because as each cylinder fires it imparts a slight twist to the crankshaft, rather like winding a spring. This sets up a torsional vibration that eventually snaps the crankshaft. The Wright brothers did not dare to run their new six-cylinder engine at the high revolutions at which this vibration set in. It was left for others to solve this problem.

One way to avoid a long crankshaft is to use a short four-throw crankshaft with a bank of four cylinders, then add an extra bank of four cylinders and pistons at 90° to the first bank, driving the same crankshaft and sharing big-end journals. And this is exactly what the French engineer Léon Levavasseur did when he invented and patented the V8 engine, the elegantly built and named Antoinette 8V.

Levavasseur was a fine arts student who clearly had an eye for aesthetics. He switched to engineering and started building petrol engines for motorboats. His V-shape motors fitted neatly into the bilges of a boat, as V-shaped steam engines had before them. In 1903 he persuaded the French industrialist Jules Gastambide that his engines would be light and powerful enough for speedboats and for the aircraft that were clearly on the horizon; furthermore, he suggested that the engine should be named after his patron's daughter: Antoinette.*

Levavasseur had patented a fuel-injected 8-litre V8 engine, a machine now so identified with the American muscle-car culture of the Sixties that it might come as a surprise that it was invented by a Frenchman. It was rather more beautiful than the average Detroit cast-iron lump; for a start each cylinder was surrounded by a thin water jacket made of electrolytically deposited copper, and these contained water that was turned to steam by the heat of the cylinders. This rose through copper pipes and was evaporatively cooled back to liquid water. The fuel was injected into each inlet port through a decidedly modern-looking stack of inlet pipes. When all the copper was polished this was a lovely piece of sculpture. The power, though, was only 50 hp, a figure that could be achieved today by a motorbike of only a quarter of 1 litre: 250 cc. By 1904 most of the prize-winning speedboats in Europe were running the Antoinette 8V engine.

Aircraft designers need lightness as well as power, though, and the specific power was 4.6 lb per hp, half that of the Manly engine. The Antoinette 8V was fitted into the first French aircraft, *14b* (*Quatorze-bis*), built by the Brazilian aviation pioneer Alberto Santos-Dumont. This aeroplane refused to take off with its first engine, and so an Antoinette 8V was fitted instead.

On 13 September 1906 at the Bois de Boulogne, *Quatorze-bis* was just able to take off, and it flew for around 5 metres (16 feet) and reached a dizzying altitude of 70 cm (28 in). Many still claim that this was the world's first true powered flight, as the aircraft took off and landed on its own undercarriage wheels, whereas the

---

* In the same year the daughter of the owner of the Daimler Motors corporation gave her name to the cars: Mercedes. Hitler perhaps never realised that his favourite Mercedes-Benz parade car was named after the granddaughter of a rabbi.

Wright Flyer took off from a launch rail, leaving a wheeled dolly on the ground.

This beauty of an engine was definitely the first to power an aeroplane flight in Britain, though, and this happened at Farnborough in 1908. But it was still too heavy for its limited power.

At this point we should briefly consider the arguments between air and water cooling, a battle between wind and water which was later epitomised by the competition between the Bristol Hercules air-cooled radial engines and the Rolls-Royce Merlin water-cooled engines of the Second World War. As we know, internal combustion engines have a vast amount of waste heat to dispose of, particularly from inside the cylinder head where most of that combustion is happening. For a small single-cylindered motorbike engine, simple cooling fins all around the cylinder and head will suffice, as you can check yourself the next time you see a suitable bike parked up.

Weight is the first consideration for aero-engine designers, and so it is natural for them to be attracted to air for cooling; after all it is cheap, light and easily obtainable. The fact that you need four thousand times more air volume than water volume to remove the same amount of heat doesn't matter: you can suck it up and eject it as you fly along. You will be carrying less weight around if you let air do the cooling.

There were other advantages of air cooling that soon became apparent as the nascent aero industry arose and the first decade of the century gave way to the second. Although the first aero engines were water-cooled, the most popular early aero engines were the air-cooled French Gnome rotary engines, designed and built by the French brothers Louis and Laurent Seguin. Based on a German single-cylinder design, their purpose-built aircraft engine had seven cylinders and produced the same 50 horsepower as the Antoinette 8V, but it weighed 44 lb (20 kg) less. It was launched in 1909 as the rotary Gnome Omega, and it was so utterly different from the V8 Antoinette that it is hard to believe that they shared anything at all in common.

For a start, instead of having stationary cylinders and a revolving crankshaft, the Gnome had a stationary crankshaft fixed to the aircraft, and the crankcase, cylinders and propeller all whirled around like a demented Catherine wheel. As a result the finned cylinders and cylinder heads were efficiently cooled by churning through the air. An amateur mechanic examining a Gnome rotary for the first time might be foxed by seeing only a single exhaust valve at the top of each cylinder. Where was the inlet? The secret was that the inlet valve was hidden in the crown of the piston, and the fuel and air mixture was inhaled through the hollow crankshaft into the crankcase and up through this valve. The Gnome was delicately made – each cylinder barrel had a wall thickness of only 1.5 mm – and yet it had been machined out of a solid steel billet, fins and all; a huge machining job. Like the Wright Flyer's engine,* the Gnome had no throttle, and if the pilot wanted fewer horses all he could do was to switch the ignition off with a joystick-mounted control. The torque reversals produced by this technique exerted terrific strains on the aircraft.

These rotaries dominated the early aviation scene partly because they were light, and this was a by-product of the circumferential layout of the cylinders, which saved a lot of crankcase weight. They were also smooth-running, well balanced, and suffered no vibration. Vibrating engines quite literally shook those early aircraft

---

* The Wrights didn't fit a throttle to their engines until 1913. A butterfly throttle restricts the air/fuel mixture to the cylinders, and is usually a centrally pivoted metal disc that allows the mixture to flow to the cylinders.

apart. Watching a vibratory engine running on the ground is quite a sight; the whole machine seems to be a blur.

There was a further advantage of the air-cooled engine: in wartime a bullet or piece of shrapnel might break off a cooling fin without affecting the running, whereas if a water jacket, pipeline or radiator was punctured it would stop the engine in minutes. They were so popular that more than 1,700 Gnomes were built across Europe.

These rotary engines were memorable in many ways. Apart from all the ironmongery spinning around, which would disconcert a car-owner, the rotary also flung castor oil around in a half-burned spray. This stuff has the most delightful, evocative smell, but after breathing it in for a while pilots found they were attacked by violent diarrhoea, an affliction difficult to cope with at altitude in a cramped cockpit wearing tight-fitting breeches.

There were disadvantages of air cooling that emerged as the number of cylinders multiplied. The last cylinders in an in-line row would be shrouded by the others and wouldn't feel enough of the cooling draught. They would start to grow too hot, and shortly afterwards a piston would seize inside a cylinder and the whole thing would then clank to a halt. There were disadvantages, too, of rotary engines. The weight of the whirling mass of metal resulted in gyroscopic precessions which had a dangerous effect on the aircraft's manoeuvrability – imagine trying to steer a bicycle with a spinning flywheel attached to the handlebars. As rotary engines grew more powerful and heavier this effect became so pronounced that the aircraft fitted with them became almost unmanageable.

In 1908 Fiat in Italy produced its delectable SA8/75, a lightweight air-cooled V8 aiming at a specific power of 3.3 lb/hp and in fact achieving a 150 lb engine delivering 50 horsepower. This was a featherweight delight, with gearwheels like cobwebs, slender valve-rockers, and even a hollow camshaft to save vital weight. Sadly it suffered from overheating like so many in-line air-cooled engines, partly because the Fiat engineers insisted on discharging the hot exhaust gases into the cooling airflow. Renault also produced an air-cooled V8 of 80 hp and solved the overheating problem by the use of an excessively rich fuel mixture: the petrol literally cooled it from within. It was more reliable but inefficient.

The French were at the forefront of this new technology; the little 20 hp Anzani that Blériot used to cross the English Channel in 1909 was built in Paris. The British were next to get in on the act, with Gustavus Green designing a 60 hp aero engine. Germany was relatively slow to get going, but then ran national competitions which were successful in producing good designs. On the Kaiser's birthday in 1912 it was announced that he was to award a prize for the best German aero engine, and this was won by a Mercedes engine which embodied a construction technique they had copied from a French Panhard engine of 1903 and the Antoinette 8V. It consisted of separate steel cylinders surrounded by welded-on cooling water jackets, and the engine was upside down, with inverted cylinders hanging below the crankcase.

Other European nations realised the importance of the new prime mover and hundreds of companies started making internal combustion piston engines of all configurations. All at once the steam engine looked heavy, dirty and outmoded.

Why *did* modern science arise in Europe? What was it about Europe that made the pace of progress so furious? How, for example, did powered flight advance from the Wright Flyer to the Spitfire in just 30 years? The answer goes back to Galileo. Modern science developed at the time of Galileo in the late Renaissance, namely the application of hypotheses to Nature, the use of the experimental method and the acceptance of a mechanical model of reality. Hypotheses of the medieval past tended to be vague ('God created everything') and numbers were manipulated in number mysticism a priori instead of being used to measure experiments a posteriori. For all his inventive genius, Leonardo da Vinci still lived in the old world, but by his mathematisation of hypotheses about the universe Galileo broke through the walls and shaped a new world.

Europe was ready for it. The Enlightenment incorporated science and reason, and these intellectual shifts made the British culture, in particular, receptive to new mechanical and financial inventions. The first Industrial Revolution began in Britain with textile spinning, steam power, iron making and machine tools at the forefront, all capitalised by central banks and stock markets and protected by patents until this small island off the coast of Europe was so wealthy that it acquired the largest empire the world had ever seen. Once the northern cities where the wealth was generated echoed with the tramp of thousands of boots into the mills and factories every morning, and streets were lined with fine new stone buildings along the Mersey, the Clyde and the Tyne. Now those same mills stand empty with the eyeless gaze of smashed panes, and charity shops line the streets where once they stamped 'Made in England'.

The second Industrial Revolution at the beginning of the twentieth century is our subject: mass production on assembly lines, new steel-making processes, the rise of the internal combustion engine and the invention of cars and aircraft. Intense competition arose in Belgium, Germany and France, and they raced to industrialise at the same pace. Some prophesied war. The stage is now set for the birth of an extraordinary company: Rolls-Royce, the maker of the Merlin.

CHAPTER FIVE

When Mr Rolls met Mr Royce

Rolls-Royce is perhaps the most prestigious manufacturer's name of all. The genius of Henry Royce and the flair of Charles Rolls together with their determination to make the best car in the world resulted in the immortal 40/50 Silver Ghost, whose engine led them on to the manufacture of aircraft engines. That expertise led to the subject of this book, the Rolls-Royce Merlin.

There have been many books written about Rolls-Royce over the years, at least 100 or so about the cars or aero engines or both, and dozens that devote thousands of words to biographies of the leading personalities. But few seem to get to the nub of the matter: what exactly was so special about this company and its products?

I have great respect for the old British engineers. I can just remember them, serious men of the middle class, adept with micrometer and slide rule. Often classically schooled, they named their companies *Acme, Invicta, Phoenix, Vulcan.* They may have done this because they were looked down upon by the literary and administrative elite as being 'trade' (compare that with the situation in Germany, where a fully trained engineer earned the right

Charles Stewart Rolls at the wheel, *circa* 1905. 'Charlie Rolls was the strangest of men and one of the most loveable.'

to the honorific 'Doktor'). The corrosive effect of Britain's old caste system held back our finest scientists and engineers, as C. P. Snow, a chemist and novelist, lamented in *The Two Cultures and the Scientific Revolution*.[1] Victorian schooling had overemphasised the humanities in the form of Latin and Greek at the expense of scientific education.

Alan Turing's wartime work at Bletchley Park on computing led him to crack German Navy Enigma machine codes, and he is now regarded as the father of theoretical computer science and artificial intelligence. However, he struggled to be accepted at Sherborne, his public school. His headmaster wrote: 'If he is to stay at Public School, he must aim at becoming educated. If he is to be solely a Scientific Specialist, he is wasting his time at a Public School.'[2]

For this ignorant attitude we can probably thank Thomas Arnold, the headmaster of Rugby school and an immensely influential educationalist, who wrote:

> rather than have [physical science] the principal thing in my son's mind, I would gladly have him think that the sun went round the earth, and that the stars were so many spangles set in the bright blue firmament. Surely the one thing needful for a Christian and an Englishman to study is Christian and moral and political philosophy.[3]

This medieval outlook led to the British political and administrative elites being deprived of vital preparation for managing the scientific world, as we will see in that most scientific of wars: the Second World War.

And Rolls-Royce was the most scientific of British engineering companies.

* * *

When Mr Rolls met Mr Royce they each found the man the other had been looking for. The story of how they met is a *locus classicus* in the history books, on hotel plaques and over 11,000 webpages: how electrical engineer Henry Edmunds accompanied Charles Rolls to Manchester on a train with a dining car on the morning of 4 May 1904. They got off the train and met Henry Royce in the Midland Hotel. Manchester tour guides still stand outside the Midland Hotel, point to the two plaques and repeat this story: 'Here's where Rolls met Royce.' And every year on 4 May Rolls-Royce enthusiasts celebrate the event at the hotel.

But it isn't true. What Edmunds actually wrote was this: 'I remember we went to the Great Central Hotel at Manchester and lunched together. I think both men took to each other at first sight, and they eagerly discussed the prospects and requirements of the motor industry which was then in its earliest infancy.'

There was no Great Central Hotel in Manchester. Neither Rolls nor Royce mentioned the meeting. And the assumption of historians has been that the men left the train at the Central Station and walked into the hotel next door, which Edmunds must have mistakenly called 'the Great Central'. However, as the author Ed Glinert points out, there was no train with a dining car arriving in time for lunch that week that stopped at Central Station. The only such train they could have caught was one that stopped at London Road (now Piccadilly) Station, which was the Great Central Railway's Manchester headquarters, not Central Station. Glinert writes:

I consulted Kelly's Post Office Directory for 1904. First, I verified that under 'Hotels' there was no Great Central Hotel. Then I turned to the entries for the surrounding streets. Under 'London Road' there was the usual Edwardian paraphernalia – tobacconist, blacksmith, draper, ostler, French polisher – then there was the entry for 'London Road station', an entrance

from the street below the station, and right bang next door 'the
Great Central Refreshment Rooms' – just the kind of smart, no
nonsense place two hungry men, with no time to waste, might
meet a third to do a bit of business.[4]

Does this kind of nit-picking matter? Well, yes. If this story, central
to Rolls-Royce mythology, is untrue, how much else do we have to
approach with caution? And it's strange to think of how many
worthies in Rolls-Royce motor cars have been going to the wrong
venue for years.

The fact that the men were pleased to meet each other is beyond
doubt. Rolls said afterwards: 'I could not find any English-made
car I really liked … eventually, however, I was fortunate enough to
make the acquaintance of Mr. Royce and in him I found the man
I had been looking for for years.' And Rolls commemorated that
first meeting by sending Royce a silver Vesta case engraved with
the date: May 4th, 1904, the conception of the company.

   Frederick Henry Royce (Henry) was 14 years older than his new
partner. He was born in 1863 in the village of Alwalton near
Peterborough. The Royce family were millers and lived at one time
in South Luffenham, the village in which I grew up, and his mother
Mary King came from North Luffenham. We knew her family, and
I know this as a deeply rural part of inner England: the country of
the Northamptonshire peasant poet John Clare. Henry Royce's
father James was also a miller, but suffered from the fatiguing
Hodgkin's disease and died in a poorhouse at the age of 41. Henry,
like John Clare, therefore experienced poverty from early youth,
and was sent out bird-scaring in the fields when he was still three
years old. Many years later the little scarecrow would be Baron of
Seaton, a village not far away in Rutland where his family had once
been millers.

At this time, however, the family had no money and moved down to London, where one of the jobs Henry undertook was telegram boy. Curiously enough he is said by Rolls-Royce historians to have delivered telegrams of congratulations to 35 Hill Street, Berkeley Square, in the Mayfair district of London on the occasion of the Hon. Charles Stewart Rolls's birth on 27 August 1877. It is hard to imagine two Englishmen born further apart along the social spectrum.

Henry Royce showed an aptitude for engineering and was sponsored by his aunt to undertake a premium apprenticeship at the Great Northern Railway works at Peterborough. Working under one of the best engineers of the day, he took every opportunity to educate himself, spending his evenings studying algebra, French and electrical engineering. For a young man of the working class the profession of engineer would be an honourable occupation. But after his aunt could no longer support him he struggled to find employment in one of the periodic economic downturns. He ended up in Leeds as a toolmaker working a 54-hour week for 11 shillings, about £90 in today's values (£1.67p an hour). For several months he worked from 6 a.m. until 10 p.m. and all through Friday night, and eventually this was the kind of commitment he expected from his own workers when he started his own business. He also neglected to eat properly, and this affected his health.

Royce showed talent as an electrician and was eventually given the technical responsibility for the installation of lighting in several streets in Liverpool. However, his employers were bankrupted, and so he set up in business as F. H. Royce & Co. in Blake Street, Manchester with his fellow engineer friend Ernest Claremont, who invested £50. They lived together in a room over the workshop, rigging up hammocks and cooking their meals in an enamelling stove, a practice that may have led to the gastric problems both men suffered in later life. According to the Rolls-Royce historian Harold Nockolds:

> Their only diversion at this time was a card game called 'Grab', which appears to have been a combination of all-in wrestling and strip poker. At any rate they both wore tightly buttoned overalls when playing the game, which generally ended in their rolling around the floor like a couple of puppies.[5]

Otherwise they were working around the clock making electrical devices such as an electric doorbell kit, which was a success and paved the way to the building of dynamos, electric motors and cranes. It was about this time that Royce patented improvements to the bayonet light bulb that we still use today. He started to show his particular genius: the ability to take apart previous inventions, work out how to improve them and then build better versions. He himself said: 'Take the best that exists and make it better.' This is one of the key characteristics of the Rolls-Royce method: they didn't waste effort reinventing a wheel; instead they made it perfectly round and unbreakable.

In those days of gaslights any ship, mill, factory or office that wanted electric lighting had to buy a dynamo, and Royce's were the best. Dynamos with heavily arcing brushes would swiftly burn out, but Royce worked out a way to achieve sparkless commutation and his dynamos just kept going. This was going to become the defining quality of the Rolls-Royce magic: reliability. And one day that was going to be a lifesaver.

Royce was what we could call today a workaholic and a perfectionist: 'For many years I worked hard to keep the company going through its very difficult days of pioneering, personally keeping our few machine tools working on Saturday afternoons when men did not wish to work.'

Men arriving in the mornings would often find him asleep at the bench with his head cradled in his arms, his machines whirling emptily. Discipline was so strict in the Manchester works that a man seen loitering or joking at his machine was likely to be dismissed without notice. He would allow a man half a day off to get married. He himself admitted that he was a shy man who suffered from excessive sensitivity. He still neglected his health; a boy in the factory would be sent after him with a glass of milk and told not to come back until he had drunk it.

The profits of the company were ploughed back in, and Royce began to make the best electric winches and cranes on the world market, so much so that when the Japanese Imperial Navy bought a crane for their dockyard and copied it, they copied the Royce nameplate, too!

There was a brief period of prosperity, and Claremont and Royce married the sisters Edith and Minnie Punt. Colleagues who marry sisters do not always find happiness for either the women or the men, and the Royces eventually separated in 1912. According to one historian both Edith and Minnie were uncomfortable with the physical aspects of their relationships.[6] However, Henry and Minnie built a fine house in Knutsford in the fashionable Arts and Crafts style. It had a Royce electric generator in the back garden, giving them the first electric lights in the town. This is where Royce tinkered with cars and grew roses, his only hobby. Oddly for a Royce creation, his house had quality problems. In 2008 the owner said: 'Half of the house had to be dismantled due to subsidence and next door's house was made with a lot of the bricks.' It would be worth around £1.6 million today, an indication of Royce's rise from his humble rural beginnings as a scarecrow.

Around the turn of the century Royce had become interested in the new motor cars, and first bought a De Dion-Bouton and then a second-hand 1901 Decauville. The Continent was in the forefront of car manufacturing at that early stage, the French in particular. Perhaps surprisingly, electric cars were thought to be the future: 'P1', the first car Ferdinand Porsche made in 1898, had to carry half a ton (500 kg) of lead-acid battery. Performance was brisk but the range was short. So not much has changed in 120 years. It is slightly odd that Henry Royce, a manufacturer of electric motors, did not buy an electric Porsche; a four-motor example was ordered by the Englishman E. W. Hart in 1900. As a consummate engineer he probably realised the advantage of internal combustion engines: petrol has an energy density 60 times that of a lead-acid battery and at least eight times better than the best present-day batteries. A battery equivalent to the Silver Ghost's fuel capacity of 12 gallons would weigh a couple of tons. When its energy was exhausted it would still weigh a couple of tons, unlike an empty petrol tank. And even a ham and cheese sandwich, weight for weight, is around 30 times more energy-dense than the best lithium battery.

As for Ferdinand Porsche, he served for a while as a chauffeur to Archduke Franz Ferdinand of Austria, the crown prince of Austria whose assassination would precipitate the First World War a decade later. (The word chauffeur comes from the French for stoker, or fireman, referring to the stoking of the boiler on early steam vehicles.) In the Thirties a keen motorist named Adolf Hitler admired the 1932 Czechoslovakian Tatra T97, a small, air-cooled, rear-engined economy car, and remarked to Porsche: 'This is the car for my roads.' Porsche copied many aspects of the design and produced the Volkswagen Beetle, which was strikingly similar. Tatra launched a lawsuit, but this was simply disposed of when Germany invaded Czechoslovakia. Dr Porsche then designed tanks and joined the Nazi Party, reaching the rank of SS-Oberführer.

After the Second World War Tatra reopened the issue, and in 1965 Volkswagen paid Ringhoffer-Tatra 1,000,000 Deutsche Marks in an out-of-court settlement.

The Frenchman René Panhard was perhaps the most influential motor-car pioneer, as he laid out the shape that the petrol-fuelled car was going to follow for 100 years: it would have four wheels, with a radiator in front followed by an engine, then a clutch and a gearbox which would send power to the rear wheels. '*C'est brutal*,' Panhard admitted of his sliding-pinion crash gearbox, '*mais il marche*': 'It's brutal, but it works.'

There were few British cars of quality available due to the infamous 1865 'red flag' speed restrictions. These called for motor vehicles to restrict their speed to 4 mph in the country and 2 mph in the town (6 and 3 km/h respectively), and for a man carrying a red flag to walk in front of a vehicle pulling a trailer. The 1896 Locomotive Act removed some restrictions of the 1865 act and raised the speed to 14 mph (23 km/h). This still inhibited the growth of the motor-car industry in Britain. No less an inventor than Thomas Edison wrote in 1901:

The motor car ought to have been British. You first invented it in the 1830s. You have roads only second to those of France. You have hundreds of thousands of skilled mechanics in your midst, but you have lost your trade by the same kind of stupid legislation and prejudice that have put you back in many departments of the electrical field.[7]

Henry Royce hadn't given much thought to building his own cars until he drove the Decauville with its two-cylindered engine. This vibrated horribly and offended his perfectionist instincts. He was dissatisfied with other aspects of the Decauville, so in typical Henry Royce fashion he dismantled it, inspected the parts, made improvements and then resolved to build his own motor car. So runs another Rolls-Royce legend.

But if the Decauville was so bad, why was there so much of the French car in the genes of the Royce? The radiator looked so similar that when Charles Rolls saw the car for the first time he thought it was a product of the French factory. The rest of it looked much the same. Royce's approach was always to 'take the best and make it better', which later became the central Rolls-Royce mantra. So he set about doing exactly that: copying what worked, improving what didn't. As he said himself, inventors and pioneers rarely make any money, only those who take their ideas and make them work.

Meanwhile the end of the Boer War had caused an economic downturn, and cheaper foreign-made dynamos appeared on the market using Royce-patented designs without paying him royalties. Royce either had to make his own products more cheaply or diversify by making a new product. The first course would have been anathema to him, and he saw clearly that the motor car was the next big thing. It would be ideal for his business. His partner Ernest Claremont was more conservative and hadn't been keen on the switch to electric cranes. He became more and more disenchanted with Royce's diversifications and resisted the move into car manufacturing, especially when Royce kept borrowing men in the workshop for the more interesting business of building his first car.

'In 1904 he had produced the first Royce car: this was before he met the late Hon. C. S. Rolls,' wrote his friend Frank Lord:

The car was a 10 hp. two-cylinder, and was a revelation for its date, having properly lubricated joints to the drive shaft [instead of chains]. As he could not buy a satisfactory coil for the ignition, he designed one, fitting very large points of the purest platinum, which, although expensive in the first place, never seemed to want adjusting or cleaning. The coil itself was as nearly perfect as possible, thus from the very first making the car reliable in a part in which, with most cars, there was endless trouble.

The two friends took the Royce car on its very first run through Wales. 'During the whole three days' trial we never had a stop of any sort from any fault of the car, a pretty good performance for a car designed by a man who had never designed one before; yet only what you could expect from one designed by Mr. Royce.'[8]

Royce's building methods clearly differed from those of other manufacturers, and this helped to achieve a high level of reliability. One example is the use of bolts where other manufacturers used rivets. These are mushroom-shaped pieces of steel used to fasten heavily stressed brackets such as spring hangers to the chassis, which is the long steel frame supporting the body. A hole is drilled in each piece and a red-hot rivet is pushed through the two holes. The 'stalk' of the mushroom is then hammered over, fastening the two pieces when the rivet cools. Here is Ernest Wooler, who had served as an apprentice under Royce in 1904:

I remember Royce carefully explaining to me as a child how a hot rivet never filled a hole when it cooled. A cold rivet was punishing the metal too much. So we made taper bolts fitted perfectly in a hand-reamed hole. It is such details that explain the difference between Rolls-Royce and other cars …[9]

This story contains Royce's philosophy in a nutshell: take workmanship to the highest possible level. A nut on a tapered bolt could always be tightened if the bracket became loose. This small detail could become of vital importance. However, it would cost much more to make. One of Henry Royce's remarks that seems to define his work was this: 'The quality will remain long after the price is forgotten.'

Another example: as the first car neared completion it was noticed that the front axle (a long bar of steel carrying the wheels) was out of alignment by half an inch. One of Royce's workers bent the axle without first heating it up red-hot, but then received a stinging rebuke from Royce, denouncing his work as 'foul practice'. It would, he said, weaken the metal. The axle was scrapped.

That first car was for Royce's own personal use, and a second was built for the more sceptical Claremont. There was a third such prototype, and these three cars were named the Royce 10. They still had only two cylinders, but Royce's meticulous balancing made it virtually vibration-free. His work on the valve gear quietened what was usually a clattering cacophony, he fitted a large exhaust silencer and eliminated every other source of vibration and rattle. The result was an exceptionally smooth, quiet car that contrasted with the others on the market. This is what so seduced Charles Rolls.

It has been said that the influence of the Honourable Charles Stewart Rolls upon the company that bore his name was negligible. After all, he had clearly lost interest in the partnership just six years after his meeting with Henry Royce, and for half of that time he was keener on flying balloons and aeroplanes than on selling motor cars. Some purists call their cars 'Royces', regarding this as a truer indication of their origin.[10] However, as we will see he brought a vital quality to Rolls-Royce Ltd, a quality the company has always traded on: prestige.

When – and if – Henry Royce delivered that telegram celebrating the birth of his future business partner, he must have felt the Rolls family was a long way from his own social stratum. Three out of four people living in Britain at the time worked as cash-in-hand tradesmen like the young Henry Royce, many of them employed as factory workers, shopkeepers and mechanics. And most aristocratic families like the Rollses would have a house in the town and one in the country, employing a large domestic staff of butlers, housemaids and cooks to maintain the household. What the Hon. Rolls was going to give Royce was access to a large number of aristocratic and wealthy customers for the Best Car in the World.

The Rolls family's fortune had been made by a dairy farmer ancestor who had bought up both sides of the Old Kent Road for property development. He married shrewdly in 1767 and thus expanded his land ownership into other parts of London. His wife Sarah also brought estates in Monmouthshire which enabled him to become High Sheriff. His son, however, did what sons often do and gambled a fortune away in the 1790s. The press, with some glee, reported that a 'dashing Cow-Keeper's son in the Kent Road has, during the past summer, been pigeoned of near £60,000'. That's well over £8 million today. The family recovered their fortunes and Charles's father, John Allan Rolls, was raised to the peerage as Baron Llangattock of the Hendre. His London estates housed 60,000 of the working classes and brought in millions of pounds. The Duke and Duchess of York (later King George V and Queen Mary) stayed with Lord and Lady Llangattock at the Hendre in 1900, and Charles took them on motor-car excursions, probably the first time that the royal couple had been in a car.

At Eton Charles showed a precocious interest in engines, earning the sobriquet 'Dirty Rolls' (at least we hope it was for the engines). He decided to go in for a life as a professional engineer, and we have already noted how unusual this would be, particularly for the son of a lord. He installed a dynamo and lighting at the Hendre, his parents' house in Monmouthshire, and he demonstrated a great deal of practical ability. Later on this may have endeared him to Henry Royce. When they met, Rolls was the better qualified by education (a Cambridge BA in Mechanical Sciences) and eight years of practical experience of buying, racing and selling foreign motor cars. Royce himself had no higher-level engineering or scientific education, but nor did other pioneers: Wilhelm Maybach, Ferdinand Porsche and Ettore Bugatti. This may have set them free in the new discipline. A close friend, Moore-Brabazon of Tara, later Lord Brabazon, described Rolls at this time:

> Charlie Rolls was the strangest of men and one of the most loveable. He was tall and rather thin, and his eyes stood out of their sockets rather more than is normal. He was rather fond of a Norfolk jacket ... and always wore a very high, stiff, white collar ... While motoring he would turn his cap back to front ... Incidentally, he was a snob, too: the way he used his superb powers of salesmanship to float early Rolls-Royce cars on the aristocracy of England left every other firm an 'also ran'.[11]

Rolls was at first fascinated by the speed of cycle racing, but he soon switched to motor cars when they started to appear. He wasn't particularly academic, and he had to attend a private crammer to get him into Cambridge. However he made sure he was the first undergraduate to go up to Cambridge by car, and indeed he was probably the first owner of a car based in that town. He undertook a pioneer drive, the first from London to Cambridge, which he completed overnight with a companion in 12 hours, an average speed of 4.5 mph (7 km/h). On the way, while walking in front of the car with the obligatory red lantern, he encountered a policeman at two o'clock in the morning. To speed them up a little he invited the officer on board and they continued on their way at 20 miles an hour, the policeman holding on in terror until the engine overheated. Rolls used the Cambridge University

Engineering Laboratories to work on his cars: first a 3¾ hp Peugeot Phaeton, then two motorised tricycles; a De Dion and a Bollee. Foreign-made cars were still ahead of anything being made in Britain.

Rolls's biographer, Lord Montagu of Beaulieu, did not seem to like him much, citing monomania, meanness and a 'banana-skin-type of humour'. He recalled a jibe from a hansom cabby, reins in hand, who yelled 'Old iron!' at Rolls's car as he thundered past. 'Take it home and eat it,' retorted Rolls (quite a well-judged reply, in fact). Montagu makes a good case for him as a visionary who contributed hugely to the progress of the motor car and aeroplane in British society. Once the revolution of the motor car was channelled into the evolutionary progress of his Rolls-Royce Silver Ghost, Rolls switched his attention to aeroplanes.

As for his character, Rolls was clearly courageous and determined. It's hard to believe as we drive our air-bagged safety machines of today, but early motor cars were extremely dangerous: an accident usually meant serious injury to driver and passengers. And an aircraft accident was usually fatal. Our modern cars evolved at breakneck speed thanks to men like the young Charles Rolls. Between 1894 and 1914 cars developed from motorised horse-drawn carriages to 120 mph (193 km/h) racing machines. We might think our present-day computer industry is developing fast, but consider this: engines grew from single cylinders to 12 cylinders, from sidevalve asthmatic wheezers to double-overhead-camshaft four-valve deep-breathers, from solid axles to air suspension, two-wheel braking to four-wheel braking, and maximum speeds from 12 mph to 120 mph (19 to 193 km/h). Moore's Law states that the number of transistors in a computer doubles about every two years, and car technology during those hectic years raced along at much the same rate of progress.

The Hon. C. S. Rolls was at the forefront, participating in every car trial he could, goggles strapped to his face, cap on backwards, thundering ahead of a column of dust, screwing the maximum out of lethal machines. It's hardly surprising that he had a taste for Wagner. But he wasn't just gung-ho; his mechanical sensitivity enabled a rare relationship with his machines. Lord Montagu said of him: 'everyone who rode with Rolls testifies to the quickness of his reactions, and to the sensation that he was at one with his steed.' He won the Automobile Club Gold Medal for best amateur performance in the Thousand Mile Reliability Trial and was fastest of three British drivers in the 1905 Gordon Bennett race in the Auvergne, France, the ancestor of the Grand Prix. The engines became more powerful, and the cars faster and faster.

The rest of the technology, the suspension and the brakes, struggled to keep up with the pace and the power. Contemporary roads were nothing more than muddy cart tracks studded with tyre-puncturing horseshoe nails. One racing 1910 Vauxhall with suspension unrestrained by effective dampers encountered a stretch of road surface at such a high speed that the road-bump spacing coincided with the natural frequency of the body bouncing on the springs. The resulting resonance caused the size of the bounces to increase to such an extent that the whole car leapt up in the air, jumped over a hedge and crashed upside down into a field.

In Britain motor cars drove on the left as the Romans had before them. (The evidence for this has been found at a Roman quarry near Swindon. The left-hand set of grooves in the road leading down and away from the quarry were much deeper than the grooves leading up to it, suggesting that the heavily laden carts were leaving on the left.) There's another good reason for driving on the left, which is that when leading a horse it is done with the left hand, keeping the right hand free for a stick (or a weapon). You would also want to walk along the middle, drier part of the cambered road, and so you would probably choose to walk the horse on the left side of the road. In the year 1300, Pope Boniface VIII directed pilgrims to keep left, and Italian cities had left-hand-driving traffic until Rome made the change on 1 March 1925 and Milan on 3 August 1926.

Brakes were virtually non-existent, and Rolls had trouble with slowing down. He had graduated from Cambridge in that same year of 1898 with a degree in Mechanical and Applied Science, and started on a career in engineering. But he wasn't a designer, more of a pioneering driver and adventurer. Lord Montagu described one jaunt:

> He determined to celebrate Christmas by driving down from South Lodge to the Hendre in the Peugeot. To moderns, to embark on such an adventure in winter with primitive brakes which over-heated at the least provocation appears sheer lunacy … such pneu-matic tyres as were available were treadless and thus incapable of any degree of grip. Anti-freeze was unknown, and one thawed out one's cooling system by lighting fires under the car.

On this journey Rolls had to negotiate the steep hill down the Cotswold Edge at Birdlip, which has a gradient of 1 in 6 (16 per cent).

> Birdlip all but proved their Waterloo. The side-brake expired on the steepest part of the hill, while the foot-brake linkage bent under the constant pressure, and the pedal went flat to the floor, leaving the Peugeot careering downhill towards a light which appeared to be a vehicle in the road, but was fortunately a lantern in a cottage window. They narrowly missed heaps of stones but reached the foot of the hill in safety.[12]

(I attempted to emulate Rolls's brakeless descent of Birdlip Hill just after writing this section. It proved impossible to descend the 1.1-mile-long hill without applying the brakes to remain within the speed limit, and on my second try a wheel hit an Edwardian-sized pothole and burst the tyre. I take my cap off to him.)

Rolls's troubles weren't over. The next morning as he bent down to wind the starting handle 'the clutch pedal jammed down, and he had the humiliation of being run over by his own car. He was unhurt, but his companion was so thunderstruck that he made no attempt to arrest the Peugeot's wild career, and it rammed a dog-cart, which Rolls had to repair before going on his way'![13]

The death toll of animals killed by the new motor cars on quiet country roads was startling: Charles Rolls's tally on just one day was two dogs, a suckling pig and four chickens. On another occasion he reported that he came around a corner with his constant-speed engine and negligible brakes only to meet a butcher's cart. The frantic horse bolted and overturned the cart, 'scattering various spare parts of animals about the road'.

One can imagine that the attitude of the wealthy owners did not endear them to the rustic owners of slaughtered livestock, if indeed they bothered to stop at all. Cathcart Wason, MP decried motor cars as 'those slaughtering, stinking engines of iniquity' and in return Lord Montagu, writing in 1966, refers to the 'thick-headed drivers of horsed vehicles and stubborn cyclists'. His description of his rural fellow countrymen as 'the peasantry' is surely one of the last uses of the word in non-ironic form.

It is hard to imagine the heady excitement of motor-car driving in those pioneer days, but Kenneth Grahame captures it for us in *The Wind in the Willows*. Toad, Ratty and Mole are strolling down a country lane when:

... far behind them they heard a faint warning hum; like the drone of a distant bee. Glancing back, they saw a small cloud of dust, with a dark centre of energy, advancing on them at incredible speed, while from out the dust a faint 'Poop-poop!' wailed like an uneasy animal in pain. Hardly regarding it, they turned to resume their conversation, when in an instant (as it seemed) the peaceful scene was changed, and with a blast of wind and a whirl of sound that made them jump for the nearest ditch, it was on them! The '*Poop-poop*' rang with a brazen shout in their ears, they had a moment's glimpse of an interior of glittering plate-glass and rich morocco, and the magnificent motor-car, immense, breath-snatching, passionate, with its pilot tense and hugging his wheel, possessed all earth and air for the fraction of a second, flung an enveloping cloud of dust that blinded and enwrapped them utterly, and then dwindled to a speck in the far distance, changed back into a droning bee once more.

Ratty and Mole are stunned and outraged, but Toad decides at once to buy one of these machines:

They reached the carriage-drive of Toad Hall to find, as Badger had anticipated, a shiny new motor-car, of great size, painted a bright red (Toad's favourite colour), standing in front of the house. As they neared the door it was flung open, and Mr. Toad, arrayed in goggles, cap, gaiters, and enormous overcoat, came swaggering down the steps, drawing on his gauntleted gloves.[14]

To Kenneth Grahame the typical motor-car owner of the time was like Toad of Toad Hall, a vulgar, spoilt, loud and boastful creature who has inherited riches from his father. Motor cars changed rural England for ever, and not for the better.

* * *

After his passion for motor cars, Rolls fell in love with ballooning, a particularly Edwardian sport involving coal gas and not hot air as a lifting medium. Rolls had his usual set of aristocratic friends with him on these jaunts. Lord Montagu said of this sport: 'Even at its zenith, ballooning had some insurmountable limitations. A balloon could only start its voyage from the vicinity of the gas works, and thus participants in "balloon house-parties" had to desert the drawing room for the less salubrious atmosphere of the local gas company's premises …'[15]

One can only begin to imagine their discomfort.

'… One of the advantages of a feudal society was, of course, the ease with which gas companies acceded to the use of their facilities.' Some of the toffs landed in insalubrious surroundings, too, such as Princess Vittoria de Teano, who landed next to a gypsy encampment but was kindly invited to partake of their gin. The Princess Teano was not used to slumming it:

> … imagine her horror when on going to a reception given by the Duchess of Sutherland she was told that the honoured guest of the evening was Lina Cavalieri, the singer. Now all Rome knows that 'La Cavalieri' began her career by selling flowers at the doors of the theatres and concert halls in Rome. Donna Vittoria described how embarrassed she was; but she soon recovered her presence of mind and immediately left the house, remarking that she was not accustomed to meeting such persons. She afterward understood that King Edward, having heard of the incident, had said that she was perfectly right.[16]

This is the milieu the Hon. C. S. Rolls inhabited, and he milked it for all it was worth. On another balloon outing the press was transported in no less a vehicle than AX 201, the original Silver Ghost, which thus received the accolade of 'a splendid racing motor car' from the *Times* newspaper. Dirty Rolls always had an eye for a sale.

Charles Rolls said to Edmunds on their train journey to Manchester that he would like his name to be synonymous with excellence, like Broadwood pianos or Chubb safes. In that he succeeded beyond his wildest dreams. When the cars are spoken of today most of the public (other than the Royce purists) refer to 'the Rolls'. In 1902 he had started a car-sales company with Claude Johnson (later to be known as 'the hyphen in Rolls-Royce'). The company of C. S. Rolls & Co. would be selling only the best-quality cars to the wealthy motorist. Because of the inhibiting effects of the Locomotion Acts there were few British cars of quality – perhaps only Daimler and Napier might be recognised today. Rolls therefore sold mostly French cars to his wealthy and aristocratic friends, but they weren't particularly reliable.

One day the previous year of 1901 he was driving a French car that developed mechanical trouble. He pulled up outside a bicycle repair shop in Reading just before closing time and asked for advice and tools. The owner said he was closing up for the night and couldn't help. However, a 16-year-old young lad who worked at the shop, one Ernest Hives, approached Rolls and asked if he could help. He got to work and soon fixed the problem. Rolls was impressed and asked the lad if he would like to come to London to be his personal mechanic. Ernest ran home to ask his mother, got permission, jumped into the car and accompanied Rolls back to his home in London. This was a fortunate meeting for the whole country, because Hives was thus serving as Rolls's mechanic when Rolls met Royce two years later. He became Royce's right-hand man at Derby, and as works manager he pushed through the production of the Merlin engine during the war. In 1941 he decided 'to go all out for the gas turbine', ensuring Rolls-Royce's leading role in developing jet engines for civil and military aviation. He then became Chairman of Rolls-Royce Ltd and ultimately Lord Hives. Such was the curiously stratified yet meritocratic nature of English society at the time. What an extraordinary life, all contingent upon a fault in a car and a mother's quick decision.

When Rolls met Royce he had been looking for a British-built replacement for the French-built Panhards and Peugeots he had been selling to his acquaintances. He deplored the lack of British manufacturers and was on the lookout for a suitable car. What he found at Royce's was a revelation. Although he disliked two-cylindered engines, and had been looking for a three- or four-cylindered car, the smooth running of the two-cylindered Royce convinced him that here was a car he could sell under his own name.

The first Royce was not large by modern standards, with four solid-spoked wheels, a radiator in front with a flat top (not the later pent-roof Rolls-Royce design), a small engine, a dashboard, a round steering wheel (tiller steering had fallen out of favour) and leather bucket seats for driver and passenger.

Rolls was so enthusiastic about the Royce car that when he returned to London he is said to have hauled his business partner Claude Johnson out of bed, telling him: 'I have found the best engineer in the world!'

Arrangements were quickly made for C. S. Rolls & Company to sell the entire output of cars from Royce Ltd. Several new models were produced, a 15 hp three-cylinder, a 20 hp four-cylinder and a 30 hp six-cylinder. This last engine proved problematic, and the solution to that problem led indirectly to the reliability and long life of the Rolls-Royce Merlin engine.

Rolls had been insistent that a six-cylinder model was included in the Rolls-Royce range, as their rivals, especially Napier, were adding them to their ranges. D. Napier was a well-established precision-engineering company, but in the numerous uncritical hagiographies of Rolls-Royce the authors give Napier rather a raw deal. The Acton-based company had made cars before Rolls-Royce, and Charles Rolls had acted as riding mechanic on a racing Napier on the Paris–Toulouse–Paris race of 1900, buying a 50 hp model the following year. Napier seemed always to be the talented poorer cousins of Rolls-Royce, and it could be argued that they made the best British aero engines of their day in the shape of the Lion and the Sabre, but they never prospered in the long shadow of the Derby company. Napier had announced a six-cylinder engine in 1903, becoming the first manufacturer to make a commercially successful six. It was described as a 'remarkably smooth and flexible' car, and Napier were the first to use the expression 'The Best Car in the World' before Rolls-Royce used the slogan. Presumably the thought had crossed Charles Rolls's mind that he might like to buy the Napier company, too. Or might he have wondered if he could start a company, which like Napier, could build six-cylindered cars?

And so, at Rolls's request Henry Royce built a six-cylindered engine. It was made up of three separately cast two-cylinder units in a line above the crankshaft, which therefore had to be long. It was rather like three of Royce's two-cylinder 10 hp engines in a row, but unlike them it kept breaking its crankshaft at a particular rate of revolutions. To his consternation three of the six-cylinder engines broke their crankshafts in early 1906.

Broken crankshafts would prove to be a problem for many makers of six-cylinder engines, including Mercedes, for whom this problem would persist for some time. Other manufacturers even had to give up making their 'sixes'. But Henry Royce realised that the problem lay in treating the engine as three two-cylindered engines. As we saw with the Wright brothers' six-cylindered engine, as each cylinder fired it imparted a slight twist to the crankshaft, rather like winding a spring. Like any other spring it would resent this treatment and respond by rapidly unwinding in the other direction. Unfortunately, at certain crankshaft rotational speeds this winding and unwinding could overlap with the crankshaft's natural resonant frequency, and increase (rather like the bouncing Vauxhall we met earlier). In a six-cylindered engine these frequencies were close to the long crankshaft's natural frequency, and at certain engine speeds these torsional vibrations set up in the long, spindly crankshaft grew worse and worse until the poor thing snapped like a carrot and the engine clanked to a stop. This would be inconvenient if fitted in a car, potentially fatal in an aeroplane.

Henry Royce in his typical fashion puzzled over this matter until he realised that if he treated the engine as two of his *three*-cylindered engines in mirror image he could solve the problem and achieve smooth running. Royce arranged matters so that the piston no. 1 mirrored no. 6 – both reached the top or bottom of their stroke at the same time. In the same way no. 2 mirrored no. 5, and no. 3 mirrored no. 4. The firing order was such that there was a bang every 120° of rotation. It is worth pointing out that this was nothing new: Napier had already figured out how to make a successful six-cylindered engine. But Royce's *pièce de résistance* was a completely new invention, contrary to the usual view of him. It was the harmonic or vibration damper, which was a sort of flexibly mounted small flywheel attached to the front end of the crankshaft and which damped out vibrations at the critical speeds. Everything smoothed out. Royce made this crankshaft slipper (friction) vibration damper for his 1906 30 hp model; however, he didn't patent it.[17] Interestingly, the Wrights eventually offered a 'flexible flywheel drive' on their six, which did much the same job and suggests that the bush telegraph was working well.

* * *

When the decision was made shortly afterwards to produce just one model, the six-cylinder 40/50, Royce used the same mirror-image layout for the cylinders and made the crankshaft much beefier, with a large centre main bearing to suppress any carrot-snapping tendencies. The crankshaft had seven main bearings in total and full-pressure lubrication to all bearing surfaces. The result of all this was the world-famous smoothness and silence of the Rolls-Royce engine, which was largely due to Henry Royce's mastery of the six-cylinder engine. It was this engine that formed the basis of the Rolls-Royce Eagle aircraft engine, doubled up to a V12 layout and enlarged to over 20 litres. In his typical style he had taken the best engine of the best car in the world and made it better, in fact good enough to be an aircraft engine.

Rolls emphasised his own name in the initial advertising. The 1905 catalogue mentioned Royce Ltd as the manufacturer of the car only on page 3: 'Works, Manchester', and while there were biographies and photographs of Rolls and his partner Johnson there was no mention of Henry Royce. However, Rolls did work hard, and took Royce chassis and engines to the Paris salon in December 1904. Incidentally it may surprise modern readers to learn that car manufacturers only supplied the chassis, wheels, brakes, suspension, engine and transmission. The bodies were built by traditional horse-drawn carriage builders such as Barkers, Rolls-Royce's preferred builder. Rolls-Royce did not provide their own bodywork until early 1946.

The reviews of the 10 hp car were glowing, and the one thing they all commented on was the silent running:

… never before have I been in a car which made so little noise, vibrated so little, ran so smoothly … indeed, the conclusion I reached then and there was that the car was too silent and too ghost-like to be safe. That the engine could be set running while the car was at rest without any noise or vibration perceptible to the occupants of the car was good; that the car in motion should overtake numerous wayfarers without their giving any indication of their having heard it, so that the horn had frequently to be called into use, was almost carrying excellence too far.[18]

(Note that the car was described as 'ghost-like'. This could have been the inspiration for the later Silver Ghost label.)

A gold medal was awarded at the Paris salon. The cars were a success! Royce was delighted to hear that the prospective buyers were impressed by the silent running, and worked even harder: '… it is only fair to add that he drove no one harder than himself. In order to solve one particularly knotty problem he did not leave the works for three days and nights, his only rest being a few hours' sleep on a bench.'

Rolls worked at promoting the cars, too. When the headmaster of Eton retired, Rolls suggested to fellow old Etonians that a Rolls-Royce would be an appropriate leaving present. He was one of the founders of the Motor Volunteer Corps of the Army and proved the value of the newfangled motor cars to the British Army by taking the Duke of Connaught for a trial drive along the south coast. The Duke was one of Queen Victoria's sons and Commander-in-Chief of the British land forces: once again Rolls was using his connections. This jaunt would eventually lead to the Rolls-Royce Armoured Car, as used by Lawrence of Arabia. In March 1906 Rolls-Royce was set up as a company, and Rolls travelled to the United States to promote the cars.

However, he had an ulterior motive: to meet the Wright brothers. Rolls was by then hooked on a new craze: aeroplane flying. In 1909 he bought a Wright Flyer Model A aircraft, made more than 200 flights, and by 1910 he was the best-known aviator in the country. He made arrangements with Royce to cease his day-to-day commitments to their company. Rolls then became the first man to make a non-stop double crossing of the English Channel by plane, taking 95 minutes. He became a national hero. But he crashed his Wright Flyer at Southbourne on 12 July 1910, during the first International Aviation Meeting in Great Britain, and was killed outright – the first Briton to die in a powered aircraft accident. Henry Royce was devastated.

Charles Rolls had been a visionary. He correctly forecast the universal acceptance of the car, constantly advocating a small cheap runabout for the average family. He instinctively knew how to build the prestige brand of Rolls-Royce, encouraging Royce to go upmarket with six-cylinders and the 40/50. Once the revolution of the motor car was channelled into the evolutionary progress of his Silver Ghost, he switched his attention to aeroplanes, and he planned to build an aero engine. His prophecies for the future of the aeroplane were accurate, including the techniques of aerial bombing, which was the eventual reason for the Merlin engine. His reservations on the subject of airships were proved right, and even his biographer failed to acknowledge his influence on recreational ballooning. Sceptics might scoff at his title and his privileges, but the new cult of the internal combustion engine needed a pacesetter in Britain, and Rolls was the golden boy of both motoring and aviation. Without him Royce might never have found someone to promote his little two-cylinder cars and he would probably have returned to making electric cranes. Without Rolls the Rolls-Royce Merlin probably would never have been built, and maybe the Battle of Britain never won.

Let's give the last word on the Hon. C. S. Rolls to Lord Montagu:

In 1870, he would have been dismissed as an aristocratic dabbler, a by-product of the eccentric squirearchy; lack of funds would have frustrated him in 1930: while in our present times his birth and education alone would involve cries of 'privilege' from 'the dotty left'. The age of the internal-combustion engine needed a central figure, and in Britain this central figure was Rolls.

CHAPTER SIX

The Best Car in the World

The Rolls-Royce 40/50 'Silver Ghost' was described as the 'Best Car in the World', an epithet bestowed on it by the magazine *Autocar* in 1907:

> The running of this car at slow speeds is the smoothest thing we have experienced while for the silence the engine beneath the bonnet might be a silent sewing machine ... at whatever speed this car is being driven on its direct third, there is no engine as far as sensation goes, nor are one's auditory nerves troubled driving or standing by a fuller sound than emanates from an eight day clock. There is no realisation of driving propulsion; the feeling as the passenger sits either at the front or the back of the vehicle is one of being wafted through the landscape.

The chassis was originally called the 40/50 and the power output was a rather relaxed 48 hp from just over 7 litres of capacity (the current Rolls-Royce Phantom V12 car develops 563 hp from 6.75 litres).

The Rolls-Royce 40/50 'Silver Ghost'. Possibly the most valuable car in the world.

We have seen that the term 'horsepower' requires caution: it depends on the size of the horse. Early cars in Britain were taxed on a formula called RAC horsepower. It did not reflect the actual measured horsepower but was calculated by a formula including cylinder-bore size, number of cylinders and notional efficiency. Cars were commonly named for their taxable horsepower, such as the Austin Seven and the Riley Nine. The name 'Rolls-Royce 40/50' referred firstly to the taxable horsepower, 40, and the measured horsepower, 50. The unintended consequence of this foolish law was that manufacturers were obliged to design engines with pinchingly narrow bores and ludicrously long strokes. They were then stuck with inefficient sidevalves instead of overhead valves, which needed larger bores, setting back the design of British cars by 40 years. Exported British cars with asthmatic long-stroke engines blew up on the freeways of the USA, while short-stroke VW Beetles from Germany (which had a more intelligent fiscal law) rattled along for ever. And at one point the British led the world in diesel-engine technology, but this advantage was lost as a result of the heavy tax imposed on diesel fuel in the budget of 1938. Such was, and is, the scientific grasp of British politicians.

The gentle state of tune of the Rolls-Royce 40/50 made the engine delightfully flexible, and the ability to drive almost everywhere in top gear was of great importance to Edwardian motorists, many of whom could not manage the 'crash' gearboxes of the day and were unable to change gear on the move.

The chassis was built at first at Royce's Manchester works, then production moved to a new factory at Derby, attracted by low rates offered by the town council.

The chassis of the 40/50 retailed at £985, and a suitable body could cost a further £110 14s, a staggering total which could have bought a house. The first customer was William Arkwright of Sutton Scarsdale Hall, Derbyshire, a descendant of Richard

Arkwright, one of the founders of the Industrial Revolution. Initial sales were slow due to the high asking price. Claude Johnson realised that excellent though the new model was, it needed to be brought to public attention if the newly floated Rolls-Royce Ltd was to succeed. Johnson had an unerring eye for publicity. He used to balance a glass of water on the bonnet of the new 40/50 while the engine was taken up to 1,600 revolutions per minute, and not a drop would be spilt. He would also balance a penny on the end of the chassis and the penny would remain where it was. He then had an even better idea. He persuaded the factory to build a special 'demonstrator'. This was chassis no. 60551, the twelfth 40/50 to be made. An open-topped Roi-des-Belges (King of the Belgians) body by Barker was fitted, which was specially finished in aluminium paint with silver-plated fittings. On the dashboard was a plaque with the name that Claude Johnson had chosen: 'Silver Ghost'. This actual car, registration number AX 201, features largely in *Those Magnificent Men in their Flying Machines*. It is now considered the world's most valuable car, and is insured for around $35 million.

Silver Ghost was a name that would resound down the years in the Rolls-Royce hall of fame, a name that at first referred to that particular car but eventually became applied to all 40/50s. What's in a name? In the 1960s a new model of Rolls-Royce was about to be launched, called 'Silver Mist' to commemorate its illustrious ancestor, when the marketing department realised to their horror that 'Mist' meant 'dung' in German. The name was speedily changed to Silver Shadow. In this the car nearly joined the glorious pantheon of unfortunate car names such as the Studebaker Dictator (built in 1933), the Chevrolet Nova ('it doesn't go' in Spanish), and the Buick LaCrosse, which in Québécois slang means 'masturbator'. Not to mention the Mazda Titan Dump and the enigmatic Tarpan Honker. But 'What's in a name?' asked Juliet.

'That which we call a rose, by any other name would smell as sweet.'

Going so far upmarket meant that the only individuals who could afford a 40/50 Silver Ghost were royalty, dictators and the very wealthy. And 'behind every great fortune lies a great crime', as Balzac tells us.[1] The Roi-des-Belges, or tulip phaeton, body might have appealed to the royal market, as the style began with a 1901 Panhard commissioned by Queen Victoria's cousin Leopold II of Belgium: the Roi des Belges himself. The style of tulip-shaped seats was suggested by Leopold's mistress, Cléo de Mérode, a dancer who was aged 22 when she met the king, then aged 61. (Leopold obviously lacked the scruples of Princess Teano, who wouldn't remain in the same room as a dancer.)

Leopold founded and exploited the Congo Free State as his own personal business venture. Forced labour, punishment amputations, torture and murder were perpetrated upon the 20 million Africans under his rule, 10 million of whom died, according to Mark Twain. Joseph Conrad's novella *Heart of Darkness* (1899) describes the apocalypse that descended on the Congo under Leopold, and *Roi-des-Belges* also happened to be the name of the Belgian riverboat that Conrad commanded on the upper Congo in 1889. As the Roi-des-Belges body had no roof, the blood-soaked Leopold and his mistress du jour could be admired by the populace as he was chauffeured around the boulevards of Brussels.

Bolshevik royalty also enjoyed luxury cars. Vladimir Lenin's Rolls-Royce Silver Ghost was purchased on 11 July 1922. Because of Moscow's deep winter snows the car was fitted with caterpillar tracks at the rear and skis on the front wheels, so that the dictator could be driven from his Gorki mansion to the Kremlin. His chauffeur was Adolphe Kégresse, formerly Tsar Nicholas II's personal driver (the Tsar had two Silver Ghosts). Lenin's crimes were many; when famine swept his native Volga region in 1891, killing 400,000 peasants, he propagandised against charitable relief efforts from America because the spectacle of death might prove a 'progressive factor' in weakening the Romanovs. Stalin and Brezhnev also owned Rolls-Royces. Wherever history was being made there seemed to be a Rolls-Royce parked around the corner.

In India, the sixth Nizam of Hyderabad, then the richest man in the world, ordered a Silver Ghost state limousine with a raised throne on gold mounts with four collapsible seats for attendants.* He died at the age of 45 before it was delivered.

Claude Johnson, the marketing genius of Rolls-Royce, had asked for a new radiator to be designed to be more in keeping with the company's new upmarket image, and it proved to be another success, surviving up to the present day as the face of the company. At first sight it looks like a chrome-plated Greek temple, but there is more to it than that. Like the facets of the portico of the Parthenon it imitates, the flat faces of the Rolls-Royce radiator are slightly curved to give the illusion of straightness. There are no straight lines on the Parthenon, and even the columns are not straight along their vertical axes but swell in the middle. This 'entasis' is intended to counteract the optical effect in which columns with straight sides appear to the eye to be more slender in their middle – to have a waist. The ancient Greeks were masters of perspective. Without entasis the Rolls-Royce radiator would have the appearance of a chromium grin.

The Rolls-Royce radiator is also a triumph of iconography. 'The composition of the radiator', observed Professor Erwin Panofsky, 'sums up, as it were, twelve centuries of Anglo-Saxon preoccupations and aptitudes: it conceals an admirable piece of engineering

* This was chassis number 2117, and the car's total lifetime mileage was only 356.

behind a majestic Palladian front; but this Palladian front is surmounted by the wind-blown "Silver Lady" in whom art nouveau is infused with the spirit of unmitigated Romanticism.'[2] Panofsky traces Inigo Jones's introduction of Palladian Classicism to the facades of stately homes such as Lord Burlington's Chiswick House. The aristocrat being driven by his servant in a Rolls-Royce would feel as though he was being preceded by his own front door.

In 1910 there was a brief craze for bonnet mascots, and Johnson decided to have one designed for his classic radiator. He asked the sculptor Charles Robinson Sykes to design an ornament conveying 'the spirit of the Rolls-Royce, namely, speed with silence, absence of vibration, the mysterious harnessing of great energy and a beautiful living organism of superb grace'. The grimly pragmatic Henry Royce objected that the mascot interfered with the driver's view. Sykes had already sculpted a mascot for Lord Montagu's 1909 Silver Ghost. His model was allegedly Eleanor (Nellie) Thornton, Montagu's secretary and secret mistress, and the figurine held one finger to her lips to symbolise their secret love affair. It was dubbed *The Whisperer*, and when Sykes received his commission for *The Spirit of Ecstasy* the model was once again Thornton. Nellie had a daughter by Montagu (whom she gave up for adoption), but she was drowned with hundreds of fellow passengers in 1915 when the SS *Persia* was torpedoed by U-boat U38. Lord Montagu survived the sinking.

Claude Johnson's 'Silver Ghost' demonstrator was prepared for a crack at a world record: the non-stop reliability run. A team of drivers, including Johnson and Rolls, drove the car with the press aboard on the course of the Scottish Reliability Trials, and they broke record after record.

What was it like to drive? As you climb aboard you notice a large, cranked windscreen with a view over the bonnet which makes it look surprisingly short. The Grecian radiator constantly reminds you that you are driving a Rolls-Royce. On top of the unsupported upright steering column is a big four-spoke wheel with a polished wooden rim. On that is a control cluster with two levers, labelled Fast/Slow and Early/Late, and the Governor. The plain-speaking Henry Royce thought that 'early/late', referring to the spark timing, was more understandable than the more usual 'advance/retard'. A plate on the scuttle reads: 'Rolls-Royce Ltd., London & Manchester' and gives the car number as 551. The driver has a snake-like bulb-horn and the front passenger is also provided with a Desmo bulb-hooter, mounted outside below the left elbow. Outboard of the driver's door, and between it and the spare tyre, are the silver-plated gear and brake levers. The gear gate is unusual, as 1st is forward and left but 2nd and 3rd positions are both down and back, then with a short movement forward into the overdrive top; reverse is between bottom and top. The handbrake operates the cable-applied rear-wheel brakes, which are fairly quiet. The footbrake is little used; it works on the transmission and is likely to bind in hot weather and lock the back wheels in a skid. The accelerator pedal is to the right of the brake, unusually for those days – it was often between the clutch and brake pedals. Beside the spare wheel there is a Cowley speedometer reading from 10 to 80 mph, with a little clock above it. The engine is idling silently.

When you drive off the Ghost will perhaps feel rather lorry-like, with a heavy clutch and the odd crunch of gears, but the flexible engine soon gets you up to 30 mph, which is a comfortable cruising speed even on main roads. The ride is surprisingly good. The passenger sits high in a comfortable leather armchair and has to maintain 1 lb fuel pressure with a vertical floor-mounted bicycle-like plated air pump, watching the gauge that reads to 4 lb/square inch. This needs constant attention, or else the engine will stop. Otherwise you glide along in silence. This is the best of Edwardian motoring.

Johnson collected a Gold Medal on his non-stop reliability run. Then the Silver Ghost continued on a journey between London and Glasgow 27 more times without stopping the engine except for Sundays, when the car was locked in a garage. After 15,000 miles had passed Johnson invited the RAC scrutineers to strip the engine and chassis down and advise on what parts had to be replaced to return the car to 'as new' condition. The engine was passed as perfect, but one or two parts of the steering had slight wear, and so did the joints in the magneto drive; altogether costing £2 2s 7d (about £250 in today's terms). A private owner would not have needed to replace anything. This was a standard of reliability that far exceeded anything Rolls-Royce's competition could manage, and Johnson made sure he broadcast their success everywhere. For a while the 'Silver Ghost' was the most famous car in the world. Not everyone was an admirer, though. Laurence Pomeroy of Vauxhall described the Rolls-Royce as a triumph of workmanship over design, by which he suggested they placed too much reliance on correcting errors that other manufacturers would have avoided in the first place. This was a criticism that could also be levelled at the Rolls-Royce Merlin.

The 40/50s were also used as rally cars. The Austrian Alpine Trial was an eight-day reliability run, known as the toughest rally in Europe and involving steep mountain passes. In 1912 a Silver Ghost was privately entered by James Radley, the pilot who had first reached Charles Rolls after his fatal crash at Bournemouth. Embarrassingly, its three-speed gearbox proved inadequate for the ascent of the Katschberg Pass, as Radley's car ground to a halt and would only continue when two passengers got out and walked. The factory took this failure seriously and only one solution was possible if the car was to maintain its reputation. The Rolls-Royce perfectionism was brought into play and cars were sent out to the Alps to reconnoitre the passes. A factory team of four cars was prepared for the 1913 event with lower-ratio four-speed gearboxes with the engine power increased to 75 bhp. This time the passes were conquered:

> The Rolls-Royces came past in great style, and I am bound to say that I have never seen anything as beautiful in the way of locomotion than the way in which they flew up the pass; we all know what a fast car is like on the level; but the sight of a group of cars running up a mountain road at high speed, with a superbly easy motion to which each little variation in the surface gave the semblance of a greyhound in its stride, was inspiring to a degree.[3]

This time Radley finished first in all the stages, and the team gained six awards including the Archduke Leopold Cup. Replicas of the victorious cars were put into production and sold officially as Continental models, but they were called Alpine Eagles by chief test driver (and later Rolls-Royce Managing Director) Ernest Hives. The whole experience also made Henry Royce wary of entering into competition.

There was competition in the marketplace, though. A new sleeve-valve engine was announced by the Daimler company which was said to be even quieter than the Silver Ghost's engine. Johnson wrote: 'It is quite difficult to know how far the new Daimler valve-less engine is going to affect us. The engine is wonderfully silent.' As we have seen, poppet valves look like pennies on a stick and are pushed down by the camshaft lobes to open the inlet and exhaust ports. In contrast to poppet valves the sleeve valve is an extra liner between the piston and the cylinder, like a tin can with no ends and with cut-out holes in the sides. These uncover bigger ports by quietly sliding up and down and side to side: the action you might employ to dry the inside of a beer glass with a dish towel. The sleeve valve required less maintenance and allowed a better-shaped combustion chamber, but higher initial cost and greater oil consumption limited the sleeve valve's appeal to car manufacturers. It had a future, though, in Bristol and Napier Sabre aero engines.

CHAPTER SEVEN

Alis aquilae –

On an eagle's wings

A great deal has been said about the First World War, that most unnecessary of conflicts. Suffice to say that after Bismarck had unified Germany, Kaiser Wilhelm II became envious of the British Empire and desirous of a place in the sun. Despite (or because of) having an English mother, plus a grandmother in the form of Queen Victoria, his propaganda attacks on the British grew more and more shrill. In August 1914 Germany invaded Belgium, and the British were obliged to declare war on the aggressors. Foreign Secretary Sir Edward Grey stood at the window of his room and gazed out into the dusk. The gas lights were being lit. 'The lamps are going out all over Europe,' he observed to a friend. 'We shall not see them lit again in our lifetime.'

This resigned reaction was shared by the directors at Rolls-Royce. Their sole product in 1914 was the 40/50 Silver Ghost chassis costing £1,060 – about £130,000 at today's prices. As manufacturers of luxury cars their goods would no longer be wanted. They discharged half of their staff; employees were told to stop paying their rent, and if they were evicted they could bring

German biplanes dive out of the sun in this composite photograph.

their families to live at the factory – a suggestion difficult to imagine in a modern car plant. Quite bizarrely, though, three days after the declaration of war the board of directors took this decision: 'the Company should not avail itself of the opportunity now possibly arising of making or assembling aero-engines for the British Government.'[1] This was almost certainly due to the influence of the risk-averse Chairman, Ernest Claremont, who had also opposed cranes and motor cars in the past. The War Office had other ideas, though.

It must not be forgotten that most of the country expected that the fighting would be over by Christmas 1914. This is why volunteers rushed to join up, desperate not to miss the 'fun'. Nations were expected to run out of money within months. The British side of the war was commanded by men whose only combat experience was of defeating Afrikaner amateurs 12 years earlier at odds of 10:1, and they had no knowledge of handling masses of half-trained volunteers. The years of grinding trench warfare were never forecast, and the war to end all wars became a death struggle between military industrial nations. This war was going to be won by civilian engineers organising the manufacture of huge numbers of novel weapons, and Rolls-Royce would soon find themselves at the forefront of that conflict. This is the story of their first aero engine, the ancestor of the Merlin: the Rolls-Royce Eagle.

Much had happened since Rolls's fatal air crash of July 1910. The death of the playboy king Edward VII had ended the aristocrats' garden-party Edwardian era, when women were expected to be picturesque and couldn't vote, the rich were unashamed of being rich and the sun never set on the British Empire. The Parliament Act of 1911 had subordinated the House of Lords to the House of Commons, another step in the slow democratisation of the British constitution. The political power of the aristocracy had imperceptibly weakened, and so had their buying power. The coming war would wipe out a whole generation of the officer class. All of these factors were going to affect the profitability of Rolls-Royce, and so despite Claremont's caution the company clearly had to diversify.

Henry Royce's grief over his partner's death, his years of overwork and the neglect of his health had finally caught up with him, and he fell ill in 1911, diagnosed with bowel cancer. His habit of cooking food in the lead-glazing ovens as a young engineer in Manchester may well have exacerbated his condition. The indefatigable Claude Johnson, appalled at the prospect of losing the company's main asset, dropped everything, appointed Lord Herbert Scott[2] as his replacement in London and took Royce on an extended holiday. They drove across France and Italy and overwintered in Egypt. On their return they were driving along the Côte d'Azur, passing through Le Canadel, when Royce held up his hand and asked Johnson to stop the car. Perhaps mindful of his return to the industrial grime of Derby, he gazed over the warm Mediterranean bay and said: 'I should like to build a house here.'

The ever-solicitous Johnson at once arranged for a house to be built. It was christened Villa Mimosa. Furthermore, Johnson made sure that from then on Royce lived in the South of England in the summer and the South of France in the winter. Another cancerous bowel obstruction suffered at Le Canadel in December 1912 required a colostomy to be carried out. In the intensely private nature of the times this was never mentioned outside domestic circles. His wife found dressing the wound difficult to cope with and she also disliked travelling abroad to Le Canadel, so she disappeared from his life. His nurse Ethel Aubin kept Henry Royce alive and working effectively for a further 22 years. He would also have a team of engineers around him who would transmit his instructions daily to Derby. He would never live there again.

Before he was killed, Charles Rolls had urged Henry Royce to build an aeroplane engine. Royce was deeply upset by his death and was disinclined to have anything to do with aircraft, but the wartime government had other ideas. Rolls-Royce had a large modern factory in Derby, crucial for the war effort. They persuaded the company to tender for the manufacture of 50 aero engines to an air-cooled V8 Renault design, and eventually Rolls-Royce received and fulfilled orders for 220 Renault 80 hp engines at between £400 and £426 each (£40,000 each today). Who would have thought that Rolls-Royce once made engines for Renault? A further batch of a hundred 40/50 motor-car chassis for military communication duties were supplied to the War Office at a profit of 12½ per cent, so business was slowly looking up. However, the perfectionist Henry Royce was not impressed by the air-cooled Renaults. Once again, he wanted to build something better.

We have seen the arguments for and against air-cooled engines. One of the most reliable air-cooled rotaries that didn't overheat and seize up was the Bentley BR1. Walter Owen Bentley was the man behind the use of aluminium for pistons, an innovation copied by Henry Royce and by the designers of hundreds of millions of pistons now racing up and down across the globe. This is the one component that is visibly similar in all piston engines, but Bentley has little credit for it today. He too had been trained in a locomotive works, and like many could have deep feelings about beautiful and powerful machinery such as railway locomotives. 'The sight of one of Patrick Stirling's eight-foot singles could move me profoundly,'[3] he wrote.

Like Charles Rolls, Bentley raced French cars and tuned them for maximum power. Playing with an aluminium paperweight on his desk one day he realised that the metal would be far more suitable for pistons than the hefty cast iron used in engines hitherto. He had a set made up and his DFP car promptly took several records at Brooklands. Because his pistons were lighter and better at transferring heat the engine could rev (revolve) faster, making more horsepower, also a higher compression ratio could be used, releasing more energy from the fuel.

When war broke out Walter Bentley visited aero-engine manufacturers to explain to them the advantages of aluminium pistons. He started with Ernest Hives at Rolls-Royce, which is why the new Eagle engine eventually featured Bentley's pistons. W.O. as he was called then went on to manufacture his own aero engines, making the large rotary BR1 which was such a success in the Sopwith Camel, 'the Spitfire of the First World War'. He used aluminium for the cylinder barrels too, but the 150 hp BR1 was still so heavy that the spinning mass of cylinders made the Camel quite a handful: like its namesake the Bentley-powered Camel was a hairy beast. The torque reaction and gyroscopic precession enabled a sharp right-hand turn that could be a winner in a dogfight, as no enemy fighter could follow it.

'In the hands of a novice it displayed vicious characteristics that could make it a killer; but under the firm touch of a skilled pilot, who knew how to turn its vices to his own advantage, it was one of the most superb fighting machines ever built.'[4]

The reluctance to turn left was so pronounced that a pilot who wanted to turn 90° left would instead turn 270° right and make a circuit of it. The Royal Flying Corps pilots used to joke amongst themselves that the Bentley-powered Sopwith Camel offered the choice between a wooden cross, the Red Cross, or a Victoria Cross.

* * *

When considering previous aero-engine designs Royce was repeating the successful path he took with perfecting his car designs. He looked at the best and made it better. Poor cooling was Henry Royce's main objection to the Renault V8 engine he was making under licence. It had eight cast-iron air-cooled cylinders in two lines of four, cooled by a centrifugal fan attached to the tail-end of the crankshaft. This proved to be inadequate, and these engines had to be run on a deliberately rich fuel-and-air mixture. The fuel was actually cooling the insides of the cylinders. If the mixture was too lean the engine could be seen glowing red-hot. Cracked cylinders, burnt exhaust valves and seized engines soon followed. The Royal Aircraft Factory, a government-funded research establishment, tried to solve this by developing aluminium air-cooled cylinders (as used much later by the Porsche 911 motor cars). They solved problem after problem: the next one was thermal expansion. The tops of the cylinders were much hotter than the bottoms, and so the metal would expand more at the top than the bottom. The result was conical bores; rather confusing to a piston travelling up and down. The solution was large fins at the top graduating to little ones at the bottom, as you can confirm if you are still mentally examining that motorbike in the car park.

As a result of seeing air-cooled engines getting hot and bothered, Henry Royce decided to stick to what he knew best: water cooling. This of course is a misnomer; the hot water surrounding the cylinders and heads has to be pumped away to a radiator where the airflow cools it; therefore all engines are ultimately air-cooled. But water does a far better job at keeping the hot bits cool, as Royce had found with his supremely reliable Silver Ghost 40/50 engine. What he had to do was to surround the cylinders and heads with a jacket filled with water, then pump it away to be cooled well away from the engine. It could then be returned to repeat the cycle.

So Henry Royce decided to use his car engine as the basis of his first V12 aero engine. This was not perhaps as easy as it sounds. The disadvantage of trying to use a converted car engine was demonstrated by the Short Brothers, whose first effort at an aeroplane, Short No. 1, was exhibited at the London Aero Show in 1909. It never flew – not surprising, as its four-cylinder Nordenfelt 40 hp car engine weighed 800 lb. It couldn't even push the biplane to the end of the launching rail, and therefore joined the exclusive club of planes that refused to fly, like the Supermarine PB-1 flying boat.

Royce therefore had to increase power and add lightness. He settled on what was essentially two 40/50 Silver Ghost engines mounted on a common crankcase and sharing a crankshaft. He drew a water-cooled 12-cylinder engine in a V configuration, with 60° between the two banks of six cylinders: this is the natural angle for equal firing intervals in a V12 four-stroke. A V12 also presents a smaller frontal area to the resisting air than the circular rotary engines. The crankshaft-breaking 'exciting power pulse frequencies' we discussed earlier would be twice those of the six-cylindered car engine, and this would ensure smooth and safe running. The bore of the cylinders was the same as the 40/50 at 4½ inches, but Royce increased the stroke to 6½ inches, giving a capacity of just over 20 litres. He was aiming at 200 hp.

To add power, Royce turned to a German Daimler Mercedes DF80 racing engine, and here again we come across Henry Royce's technique of copying best-practice designs and improving upon them. He may have remembered Charles Rolls back in November 1905 saying of him: 'Mr Royce is an unprejudiced man and, although possessing originality to the very highest degree, he is not too proud to acknowledge the valuable work and experience of our friends across the Channel and to benefit thereby.'

Some of their friends across the Channel would now have cheerfully killed him, because with his usual foresight Claude

Johnson had bought a Mercedes racing car in the July of 1914.*
This was fitted with a monster of an engine: the Daimler Mercedes
DF80 aero engine. In 1912 a *Kaiserpreis* had been offered for the
best German aero engine by Prince Henry of Prussia, and the
Daimler company had entered a 7-litre six-cylindered engine with
lightweight 'upperworks'. These were forged steel cylinders closely
surrounded by lightweight water jackets, with overhead valves
seated directly in the cylinder heads operated by an overhead
camshaft. The power was 84 hp, and the whole thing weighed only
313 lb (142 kg), meaning that the crucial weight-to-power ratio
was 3.7 lb/hp. Although this engine came second to a Benz engine
in the competition, it was the one the German air service preferred,
and it became the basis for a range of derivatives which steadily
increased in power.

Henry Royce examined this engine closely and discovered an
Achilles heel: the crankshaft. In the racing car the engine ran so
roughly that a 1913 Le Mans race contender lost its bonnet due to
the vibration, and another had to be withdrawn from the 1914
Indianapolis 500-mile race because the vibrations were too severe
over 1,400 rpm. The crankshafts had a tendency to snap, and Royce
was confident that he knew how to solve that particular problem.

So, taking the best features of the German engine, Royce
replaced the heavy cast-iron integral cylinder block and head of
his 40/50 engine with Mercedes-style individual cylinders and
water jackets. He wrote on 7 December 1914: 'these should be
made as close as possible … we do not want to carry weight of
water which is no use to us …'[5] They proved to be expensive to
make and they leaked water in service, leading to overheating. He
also copied the Mercedes overhead valve design, discarding the

---

* For £1068 9s 2d; about the price of a new 7½-litre 40/50 Silver Ghost
chassis.

old-fashioned sidevalves of his 40/50 design. However, he improved the camshaft drive with his own 1912 patented friction-damped spring drive (probably the cleverest part of the engine).* Thanks to W. O. Bentley he fitted aluminium alloy pistons instead of the heavy old cast-iron variety. One of his original ideas was a reduction gear to drive the propeller at a slower speed. The propellers of the day did not like to be driven too fast as the tips started to approach the speed of sound, so if the propeller was attached directly to the crankshaft the engine was limited to run at only 1,400 rpm or so. Henry Royce knew his engine could make more horsepower if it revved higher, rather like a cyclist who changes down a gear to pedal faster. This was a curious blind spot on the German side, whose engines were forced to run slowly because of the lack of reduction gears. Royce's reduction gear was a more complex but superior epicyclic gear which saved the crankcase from the stresses set up by cruder spur gears (which look like ordinary cogs). Once again he was doing the best job he could, and it showed in the reliability of the Eagle. Many of the parts of Royce's new V12 were made of aluminium to save weight, and the engine came out at 820 lb (372 kg). The Silver Ghost 40/50 car engine of only six cylinders weighed 80 lb more and produced a fifth of the power!

Within six months of work starting on the drawings Royce had the proof that he had designed a winner. On the test stand his new engine developed an outstanding 225 hp at 1,600 rpm, more than contracted for: the War Office expected 200 hp. By the end of the war it was making 360 hp, and the vital weight-to-power ratio was 2.57 lb/hp. The engine was to be fitted in the new Handley Page

---

* 1967 Cosworth DFV Formula 1 racing engines suffered camshaft drive vibrations which were so severe the car became a blur when idling. They could have solved this by looking at the 1915 Eagle!

0/100 bomber and the FE2d fighter, and it was named 'Eagle' in November 1915.

The convention of naming the Rolls-Royce piston aero engines after birds of prey continued with the Falcon, Hawk, Kestrel, Goshawk, and of course eventually the Merlin, a small falcon once popular amongst English noblewomen who used them to hunt skylarks. It is rather endearing that these engines of war and death bore familiar names, and it does seem a whimsically English thing to do. These birds would have been familiar to the young scarecrow Henry Royce during his Northamptonshire childhood. The Rolls-Royce jet engines continued the theme, suggesting powerful flow by the choice of river names such as Welland (Royce's ancestors had a mill on the River Welland at Seaton). Later came jet-engine names such as Avon, Nene and Trent. The Italian and German manufacturers seemed happier with numbers to denote their machines, such as Fiat AS6, BMW 801, Daimler-Benz 600 and Jumo 211. This tendency survives with their motor-car names.

Owners of a 2005 Ford Focus ST2 might now point out that their car's engine also develops 225 hp, the same as the Eagle. Surely their little five-cylinder engine of only 2½ litres doesn't produce the same power as the mighty Rolls-Royce Eagle of 20 litres? Well, yes it does, and this is what is so astounding about technological development. With advancements in petrol octane, forced induction in the form of a turbo-supercharger and high revolutions, ordinary family motor cars can now achieve power outputs Henry Royce could only dream of. Back in 1915 the poor old Eagle had to run on 40-octane petrol that was little better than whale oil, and supercharging was far in the future.

* * *

It might surprise a modern reader, but in 1916, in the middle of the war, an 'English-Mercedes-Daimler' syndicate made a claim for infringements of two Daimler patents in the construction of the V12 Eagle. The Rolls-Royce directors considered this claim and sought advice from counsel, as they stood to lose £5,000 in royalties, a considerable sum. To show sensitivity about the enemy's possible commercial rights in the middle of total war seems surprising, and it took two years before the claim was thrown out and costs awarded against the German company. Their lawyer might have suggested the defence of *inter arma silent leges*: 'in times of war, the law falls silent.'

Had Henry Royce copied the Mercedes engine? No, he had copied another one. The concept of individual cylinders having water jackets built around them had been well established by the 1903 Mors that Rolls raced,[6] the Panhard of 1903 and also by another British aircraft engine, the Green C4. Overhead camshafts had been used in 1902 by the Maudslay motor car. It seems extraordinary that the directors should even consider this Daimler claim.

However, the enemy got their opportunity to study the new Eagle engine in June of 1916, when a brand-new FE2d fighter was being delivered to France by a pilot who had never been there before. He left the RAF factory at Farnborough, crossed the Channel, accidentally overflew the British lines and landed at the German-occupied airfield at Lille. The Germans were thus handed a brand-new example of the latest and most secret British aero engine, and so they could study any alleged patent infringements to their heart's delight. If that wasn't bad enough, on 1 January 1917 the pilot of a brand-new Handley Page 0/100 bomber got lost, landed to check his location and was politely informed that he was now in German-held territory for the duration of the war. This time they got two more new Eagle engines, together with the secret new bomber, which was duly demonstrated to the Kaiser. Johnson was so furious that he had a question asked in the House of Lords. There was no good answer.

Handing aircraft to the Germans became a bit of a habit due to an accident of meteorology: the prevailing winds over the front lines meant that any ailing Allied aircraft was generally blown over to the German side, whereas German ones were blown back home. On the other hand, a Fokker Eindecker E.III fighter that landed by mistake on a British aerodrome was tested and found not to have the superior performance it had been credited with. The Germans were complimentary about the Rolls-Royce Eagle; a 1918 report by one Herr Schwager described them as 'not the results of long-lasting experiments but the achievements of a thinking designer'. In this he was not quite correct, as Henry Royce's talent lay in endlessly testing his designs until they were reliable.

Those early days of air combat might seem amateurish now. No one knew if aircraft were going to be useful in warfare, and the first machines were primarily spotters, observing the fall of shot from artillery and recording troop movements. When an enemy aeroplane was encountered, potshots from revolvers or rifles might be attempted, but with poor accuracy. Early in the war the British pilot Flight Commander Louis Strange obtained an American-built Lewis machine gun and devised a crude mounting for it. On 22 November 1914, flying an Avro 504 at 7,000 feet (2,134 metres), he and his observer Lieutenant Small came across a German Aviatek two-seater spotter aircraft and manoeuvred into position for an attack. Freddy Small fired a drum of bullets from the Lewis while the German Army observer defended himself with a Borchardt target pistol. He was a good shot, because he wounded Small in the hand. However Small was still able to change the drum while Louis Strange turned the Avro to re-engage the

Aviatek. Lieutenant Small reopened fire and was rewarded by a stream of smoke from its Mercedes engine. The Aviatek started to lose height and eventually landed behind the British lines.

Louis Strange hastily landed, jumped onto a borrowed motorbike and rushed to examine the downed aircraft. The only damage seemed to be a few smashed instruments. He was informed by his amused Army comrades that the observer, who was a Prussian Guards officer, had thought the non-commissioned officer pilot had landed because he had been hit by gunfire. When he found him uninjured he knocked him to the ground and began to kick him viciously before he was restrained.

Six months later Flight Commander Louis Strange had one of the most bizarre experiences of the war. He was flying a Martinsyde biplane at 8,500 feet (2,590 metres) with a Lewis machine gun mounted on the upper wing above him. Once again he was attacking an Aviatek and once again he emptied a drum of ammunition. He turned away and reached up to change the drum. This was usually a one-handed operation but the drum was jammed, and so he wedged the control stick between his knees, stood up in the cockpit and grasped the gun with both hands. However his seat belt now slipped down around his ankles and his knees lost their grip on the stick. The Martinsyde immediately stalled and turned upside down, starting a slow spin down to the ground. The engine stopped. Louis was flung out of the cockpit but clung grimly to the drum with both hands, now dangling beneath the upturned aircraft. All he could see was the propeller and the slowly revolving French town of Menin beneath him. He vaguely wondered which part of the town he was going to crash into.

Then he grabbed a strut with one hand, then another strut with his other. He got his chin onto the wing and then:

I kept on kicking upwards behind me until at last I got one foot and then the other hooked inside the cockpit. Somehow I got the stick between my legs again, and jammed on full aileron and elevator; I do not know exactly what happened then, but the trick was done. The machine came over the right way up, and I fell off the top plane and into my seat with a bump.[7]

With a sickening lurch, the Martinsyde recovered from the spin. Louis was now surprised to find that he could hardly see over the edge of the cockpit. The seat cushion had fallen out during the spin and, more crucially, his fall had broken the frame of the seat. He was sitting on its wreckage, which was now jamming the controls. Meanwhile, the Gnome engine had restarted and he was roaring down at full throttle in a power dive towards Menin, which was now terrifyingly close. Hastily flinging bits of seat out from under him, he somehow pulled out of the dive and just avoided hitting the ground. Now, dodging German rifle fire, Strange flew back over the lines to his home airfield and landed. Next day he was called to see his commanding officer and reprimanded for causing 'unnecessary damage to the instruments and seat'.

The Fokker Scourge was worrying the Allies from August 1915 to early 1916. The Imperial German Flying Corps, equipped with new Fokker Eindecker monoplane fighters, gained an advantage over the Royal Flying Corps and the French Aéronautique Militaire. The Fokkers had machine guns that could fire through the propeller arc due to the invention of synchroniser gear, which ensured that bullets went between the propeller blades instead of chopping them off. This meant the whole aircraft could be aimed at the target instead of hosepiping a gun on a swivel. The technique was to fly high, dive out of the sun, line up the gunsights and fire a long burst, continuing the dive until out of range. The only aircraft the Allies had that could compete were the Airco DH2 and the Nieuport 11, both of which were fitted with a Gnome nine-cylinder air-cooled rotary of only 110 hp. This had an unfortunate habit of shedding cylinders as it flew along, but at least in the DH2 the pilot was sitting in front of any flying ironmongery: the engine was behind him. The other problem was that it used copious amounts of castor oil, which the revolving cylinders flung around with abandon. As we have seen, the laxative effect of this on pilots was well known, and considered even worse than the laxative effect of sighting a flight of Fokkers.

A flavour of the battle in the skies is provided by Cecil Lewis, a British pilot who later became one of the founders of the BBC. In his classic First World War book *Sagittarius Rising* he describes the Morane scout aeroplane:

I had a look over her, and the more I saw of her the less I liked her. It was certainly not love at first sight … the elevator was as sensitive as a gold balance; the least movement stood you on your head or on your tail. You couldn't leave the machine to its own devices for a moment … the Morane really was a death trap … Subsequently I flew every machine used by the Air Force during the war. They were all child's play after the Morane but I did come to love the Morane as I loved no other aeroplane.

He was flying one of these aircraft over the lines early in the morning on the first day on the Somme, 1 July 1916. Below him were men such as George Mallory of Everest, who was a gunnery officer, and another future Everest climber, my cousin Howard Somervell, a surgeon, together with comrades who were about to suffer 19,240 fatalities on that dreadful day.[8] The Allies gained just three square miles of territory. Lewis was ordered to report on British troop movements and witnessed the blowing up of the mines under La Boiselle:

> the earth heaved and flashed, a tremendous and magnificent column rose up into the sky. There was an ear-splitting roar, drowning all the guns and flinging the machine sideways in the repercussing air. The earthly column rose, higher and higher to almost four thousand feet.[9]

Lewis's aircraft was hit by lumps of debris thrown up by the explosion and the bang was heard in London. He also witnessed 9-inch howitzer shells turning over in flight at 8,000 feet (2,438 metres) before descending onto the target. When confronted by a Fokker, Lewis wrote: 'Hearsay and a few lucky encounters had made the machine respected, not to say dreaded by the slow, unwieldy machines then used by us for Artillery Observation and Offensive Patrols.'

The arrival of the 225 hp Rolls-Royce Eagle improved the Allied mood, as their pilots were becoming understandably reluctant to engage in combat. It was fitted to the RAF F.2 fighter and the Handley Page 0/100 patrol bomber. Henry Ricardo wrote of the Eagle:

> This engine which was developed by Rolls-Royce during the war, proved to be undoubtedly the most satisfactory and reliable engine in the hands of the Allies, and was of great value, not only on account of its magnificent performance, but perhaps even more because of its encouraging effect on the morale of Allied pilots. Official records compiled in France during the war show that the average number of hours flown by these engines between overhauls was 103.2, or very nearly double that of any other aero engine used in British service. This engine, also, is of

interest because it is at once probably the most complicated and quite the most reliable engine yet built for aircraft.[10]

Henry Royce did an outstanding job, bearing in mind that he had a debilitating medical condition and was 200 miles away from his factory. When the need for a lower-powered engine became clear he didn't simply make a cheaper V8 using the same cylinders, but insisted on scaling the big Eagle down 8/9 linearly. All the good reasons for a V12 still obtained. This smaller engine became the Falcon and was particularly successful in the Bristol fighter. The engine became renowned for reliability, often flying over 100 hours between overhauls; three times as long as other engines, with twice as many cylinders. It was celebrated in 'The Ballad of a Bristol':

But none of 'em knows the secret
Of making my heart rejoice,
Like a well-rigged Bristol Fighter
With a two-six five Rolls-Royce.[11]

Italy was on the Allied side in the First World War, and their top fighter ace was Francesco Baracca, with 34 victories credited to his name. He had started his military career with the Piemonte Reale Cavalry, whose emblem of a prancing stallion he adopted for the side of his SPAD S.VII fighter. When Baracca was killed, Enzo Ferrari met with Baracca's mother, the Countess Paolina Biancoli, and she asked him to use the Cavallino Rampante on his racing cars, since it would bring him good luck. That prancing-horse insignia is still used on Ferrari cars, another example of the connection between aero engineering and the car industry.

* * *

We should not forget the Rolls-Royce Hawk, another successful Rolls-Royce engine of the First World War. As we have seen, Royce was unimpressed by the air-cooled Renault V8 he had been making under licence, and there was a requirement for a reliable engine of around 75 hp. So in 1914 he had taken his six-cylinder 40/50 Silver Ghost engine's bottom end, increased the stroke and topped it off with separate cylinders and cylinder head, just as he did for the Eagle. The Hawk was essentially one half of an Eagle, but without the reduction gear. There was huge pressure on Rolls-Royce from the Ministry of Munitions to produce more engines for the war effort, so unusually Royce subcontracted this particular engine out to another manufacturer: the 32-year-old Roy Fedden of Brazil Straker in Bristol. Only 204 Hawk engines were made, whereas 4,681 Eagles were built. The six-cylinder Hawk was used on the SSZ-class coastal patrol airships which helped the war effort by deterring U-boat submarines from attacking coastal convoys, and the Hawk gained a great reputation for reliability, with flights of over 50 hours achieved. The purr of Rolls-Royce multiple cylinders filled the air.

The Ministry of Munitions and the Air Board were panicking about the slow rate of aero-engine production. They exerted great pressure on Rolls-Royce to allow other companies to make their engines under licence, but Claude Johnson was adamant that Rolls-Royce quality would not be upheld, and said the plan would 'yield nothing but mountains of scrap'. Was he right?

Rolls-Royce was not a mass-production company, and until 1914 it had not made one aero engine, so perhaps he felt protective of the Rolls-Royce reputation. And yet in the next war the Merlin was made in the thousands under licence by Packard and Ford. This stubbornness almost led to Rolls-Royce being nationalised, and thus forced to let others make their engines. The justifiable frustration being felt by the officials led to a disastrous decision. It shows the consequences of an engine builder NOT doing a proper job:

> The ABC engine, the Dragonfly, was a piece of rampant opportunism that might have lost us the war had the war lasted much longer; for the government were completely seduced by its design and the salesmanship of its designer Granville Bradshaw.[12]

Thus spake L. J. K. Setright, the Old Testament prophet of motor engineers. Granville Bradshaw was a considerably better salesman than designer, and he managed in early 1918 to persuade Sir William Weir, the Director of Aeronautical Supplies, to 'cancel all the others'; i.e. stop production of all other well-tried and tested engines apart from the Rolls-Royces and the Siddeley Puma. Weir made the disastrous decision to place large orders for the untried Dragonfly, with 11,500 engines contracted from 13 suppliers by June 1918.

The Dragonfly looked good – an air-cooled radial with shiny copper-plated fins for supposedly better cooling. When it finally ran it weighed 66 lb (30 kg) more than promised and developed 45 fewer horsepower. It proved to be one of the worst cooled engines ever built, growing so hot that the cylinder heads glowed red and charred the wooden propeller. Worse was to follow: it turned out that it was designed to run at the torsional resonance frequency of its own crankshaft, causing severe vibrations that caused it to break up after just a few hours in the air. Sir William's decision and Bradshaw's cupidity could have lost the British the war in the air.

* * *

The main production of the Eagle was during 1918, and this was because in the previous year the War Cabinet had at last decided to retaliate against German bombing of English cities and to start a bombing campaign of their own. They ordered nearly 700 Handley Page twin-Eagle-engined 0/400 bombers and 255 Handley Page V/1500 four-Eagle-engined bombers. This seems now a curiously late decision, largely made to placate public opinion.

Aerial bombing had first been employed during the Italo-Turkish war of 1911. On 1 November the first-ever aerial bomb was dropped by Sottotenente Giulio Gavotti on Turkish troops in Libya from a German Etrich Taube aircraft. In return the Turks, lacking anti-aircraft weapons, were the first to shoot down an aeroplane by rifle fire. Early in the First World War, on 24 August 1914 a German Zeppelin airship bombed the Belgian city of Antwerp and killed ten civilians. Later in January 1915 the Germans began indiscriminate aerial bombing of England by Zeppelin airships and Gotha bombers. Great Yarmouth, Sheringham and King's Lynn were the first targets of the Zeppelins, and on 30 May 1915 London was bombed for the first time. The novelist D. H. Lawrence watched the raid:

> Then we saw the Zeppelin above us, just ahead, amid a gleam-ing of clouds: high up, like a bright golden finger, quite small …
> Then there was flashes near the ground – and the shaking noise. It was like Milton: then there was war in heaven … I cannot get over it, that the moon is not Queen of the sky by night, and the stars the lesser lights. It seems the Zeppelin is in the zenith of the night, golden like a moon, having taken control of the sky; and the bursting shells are the lesser lights.[13]

Five hundred and fifty-seven civilians were killed, and British public opinion demanded that Something Must Be Done. These memories persisted into the Thirties and drove the orders for Merlin-engined fighters and bombers. Meanwhile British bombers were built as fast as they could be turned out.

Initially the British bombing campaign was small, with a maximum of 40 bombers carrying half a ton of munitions to drop on the Ruhr district. They were forced to fly at night and thus accuracy was extremely poor.

The first three huge Handley Page V/1500 aircraft were delivered to a squadron in Norfolk near the end of the war in November 1918, and the plan was to bomb Berlin. Just as the aircraft were preparing to leave on the mission an aircrewman ran out waving his arms and stopped them: the Armistice had just been declared.

The whole exercise had been a huge waste of money. For the same price as a four-engined bomber for offence the British could have bought nine single-seater fighter aircraft for defence. As Derek Taulbut, the Rolls-Royce historian, remarked: 'Regarding the contribution of this expensive "strategic" air campaign by the RAF towards winning the war, it is doubtful if it had any effect at all.'

The Handley Page Type 0/100, for which the Eagle was designed, was supposed to find and attack warships at sea, and was only used once for this purpose, damaging just one destroyer. However, on the plus side the RAF F.2 fighter was successful in countering the German Albatros D.1 fighter, and the Eagle-powered Felixstowe flying boats were a success on anti-submarine patrols and so did contribute to the war effort.

If Handley Page V/1500 bombers seem a waste of national resources, consider the sorry history of the American Liberty aero engine. A month after the USA entered the First World War in May

1917 the Aircraft Production Board decided to build an engine that was cheaper and more powerful than anything the Europeans could make. They brought two top car-engine designers together at the Willard Hotel in Washington and told them to stay in their rooms until they had produced a set of basic drawings.

They shouldn't have left the keys to the minibar, because the design was a disaster. A V12 of 27 litres (the same as a Merlin), it had an angle of only 45° between banks instead of the accepted 60. The separate cylinders and exposed valve gear looked distinctly old-fashioned, and there was no propeller reduction gear, so the engine couldn't rev fast enough to make any power. Despite all this, contracts for no fewer than 22,500 Liberty engines were issued to six Detroit motor-car magnates who stood to make millions of dollars. But when they tried to manufacture this monstrosity, mass-production techniques failed, as innumerable changes had to be made on the line to problems such as leaky water jackets and bearing failures. The Americans had to turn to Bugatti in Europe to help with the latter, and we still use to this day the steel shell bearings with a thin layer of babbitt (a soft alloy of tin, copper and lead used to line bearings) that Bugatti developed. When the Liberty finally ran it had problems with detonation, and the engine never really became reliable until after the war finished. Only a handful saw service in France after the whole Liberty programme – which cost around $35 million dollars ($0.64 billion today) – came to an end.

A sad footnote to this tale of woe was the fitment of a six-cylinder version of the Liberty into the Christmas Bullet Flexible Aeroplane, possibly the worst plane ever built. This was also known as the Christmas Strutless Biplane, as the cantilevered wings stuck out of the sides without any struts or wires between the upper and lower wings to support them. The designer, Dr William Christmas, was a charlatan with no experience of building aircraft. He convinced his backers that his aeroplane was part of an audacious plan to kidnap Kaiser Wilhelm II. In fact it was a death trap. The test pilot, Cuthbert Mills, invited his mother to watch the first flight, and she could only watch in horror as a wing peeled off and her son spiralled to his death. The borrowed Liberty engine was also destroyed. Completely unrepentant, Christmas built another Bullet and exhibited it at Madison Square Garden. On the test flight the second Bullet crashed, this time killing the test pilot Lieutenant Allington Joyce Jolly. Dr Christmas ended up as the vice president of a Miami-based real estate company, and died a rich man at 94.

The vast expenditure of money, materiel and lives during the First World War achieved very little. If anyone is in doubt about the futility of war – the expense, the inhumanity and the horror of machine-gunning young men, killing pilots and bombing civilians – examine the history of early aviation.

A WWI pilot with his Sopwith Camel, 'The Aircraft That Wouldn't Turn Left'.

CHAPTER EIGHT

A Rolls in the desert was above rubies

When war came in July 1914, development of the Rolls-Royce 40/50 car ceased, and all available Silver Ghost Alpine chassis were requisitioned. Some came from the production line and some were recalled from the coachbuilders. They were fitted with stronger road springs, a strengthened back axle and double-spoked twin-rimmed rear wheels to take the weight of a boilerplate-armoured body and a rotating gun turret 5 feet (1.5 metres) in diameter containing a water-cooled Vickers-Maxim machine gun: four tons in all. The 40/50 engine was largely untouched, and as we have seen was soon to form the basis of the first Rolls-Royce aero engine. This would lead to the development of the Merlin.

Squadron No. 2 of Rolls-Royces was sent to Egypt to help the Arabs' revolt against the Turks who occupied their ancestral lands, and one of the drivers was named Sam Rolls, a motor mechanic (and no relation to the Hon. C. S. Rolls). He left us a book about his experiences: *Steel Chariots in the Desert*. The Ottoman Empire fought the British and French allies in the Middle Eastern theatre during the First World War, as Turkey had allied herself with Germany. The object of the Arab Revolt was to create a unified

Lawrence of Arabia, before upgrading to a Rolls-Royce armoured car.

Arab state stretching from Aden in Yemen in the south to Aleppo in Syria in the north. The British and French promised to support this ambition, a promise that was secretly broken by the 1916 Sykes–Picot Agreement which divided up Arabia into spheres of influence controlled by Britain, France and Russia. Meanwhile, using guerrilla tactics in a form of asymmetrical warfare, the Arabs and their British and French allies struck at the conventional Turkish forces. This pinned down Ottoman troops who would otherwise threaten the Suez Canal, Britain's vital artery to India. British and French liaison officers were sent to help their Arab allies, and one of these was Lawrence of Arabia.

Sam Rolls had been driving one of the Rolls-Royce armoured cars up the Itm Gorge, a stretch of near-impassable desert north of Akaba. He spotted a group of dishevelled Arabs mounted on richly harnessed camels. One of these knelt his camel and approached the cars.

… he hastened towards me, which struck me as strange. Looking now, for the first time, full into his eyes, I had a shock. They were steel-grey eyes, and his face was red, not coffee-coloured like the faces of other Arabs. Instead of a piercing scowl there was laughter in those eyes. As he came close I heard a soft, melodious voice, which sounded girlish in those grim surroundings, say, 'Is your captain with you?' He spoke in the cultivated Oxford manner. I dropped my cigarette in sheer astonishment.

'Who the …? What the …?' I stammered out.

He placed his hand for a moment on my shoulder.

'My name is Lawrence,' said he, 'I have come to join you.'[1]

And so began the most astounding adventure, all told in Lawrence's account of his part in the Arab Revolt: *Seven Pillars of Wisdom*. This book, written in high style, had this assessment from Winston Churchill: 'It ranks with the greatest books ever written in the English language. As a narrative of war and adventure it is unsurpassable.' However, the book had an ironic subtitle: *A Triumph*, which hints at something else. It refers to the loss of self-respect that Lawrence suffered when his friends the Arabs realised that the promises of self-rule were to be broken by the politicians. Poor Lawrence. Since his mysterious death on his beloved Brough Superior motorbike his reputation has been slowly dismembered by sceptical military historians, sexually curious biographers and salacious film-makers. They have discovered inconsistencies in his account of the Revolt, but they have also been attracted by his contradictions: supreme leadership abilities joined to imperialist doubts; a deep love for a young Arab man coupled to a masochistic sexuality; his flamboyant arrogance; his flowing Arab robes, his revolvers; followed by his extraordinary effacement of his whole identity after the war.

Steel Chariots in the Desert adds to the story of this tortured soul. Particularly that aspect of his character which has been ignored by his many biographers: his extraordinary facility with machinery, which became a kind of monastic occupation towards the end. Sam Rolls corroborated many details of *Seven Pillars of Wisdom* and he also gave us some valuable insights. He had little time for the large number of officers in the cars he serviced:

Our captain had found a good supply of brass-hats. Colonels and majors there were in plenty, and, all told, I believe there were as many officers as rank and file in the British details of the Arab expedition at this time. The people who gave orders were so numerous that it was quite a privilege to be among those who only received them.[2]

But Lawrence was different. Rolls, who became his personal driver, worshipped him: 'What a man! He took men at their true value and paid no heed to outward rank or social position, and it was this attitude that made him not only a friend and helper, but a great leader.'

He noticed something else about the man which hints at his character:

'I noticed with surprise that Lawrence's feet were entirely bare. He had not even sandals. "Good Lord!" I exclaimed, "how can you possibly walk on these sharp stones, sir?"

'"Just practice," he answered quietly.'

Sam Rolls and Lawrence travelled many miles together in the Rolls. It is possible that he was relieved to be leaving camels behind. In his first engagement, the taking of Akaba, he had called to his men and they poured down the hill, shooting as they rode. Lawrence's favourite red camel suddenly went down and he was flung to the ground. After the battle was won, his men examined the camel and found that he had accidentally shot it in the back of the head with his Colt revolver.

Lawrence was curious about the powers of acceleration and speed: 'it was plain that he enjoyed the very sight of a fast-moving machine.' There is Rolls's account of a desert race between two Rolls-Royces when they reach nearly 70 mph, not bad for four tons of boilerplate and only 80 horsepower. The engines were paragons of reliability and power. The race was won by Lawrence's understanding of the sandy surface and what lay beneath it.

Their job was to harass the Ottoman forces, and in particular to attack the Hejaz railway that ran from Damascus to Medina through the desert. By constantly blowing up the tracks and locomotives but deliberately never quite destroying the railway, they tied down battalions of Turkish soldiers who would otherwise have fought the Allies elsewhere. The Rolls-Royces were reliable, but there were breakages. Lawrence describes what happened when a spring bracket broke:

A Rolls in the desert was above rubies; and though we had been driving in these for eighteen months, not upon the polished roads of their makers' intention, but across country of the vilest, at speed, day or night, carrying a ton of goods and four or five men up, yet this was our first structural accident in the team of nine.

Rolls, the driver, our strongest and most resourceful man, the ready mechanic, whose skill and advice largely kept our cars in running order, was nearly in tears over the mishap. The knot of us, officers and men, English, Arabs and Turks, crowded round him and watched his face anxiously. As he realized that he, a private, commanded in this emergency, even the stubble on his jaw seemed to harden in sullen determination. At last he said there was just one chance. We might jack up the fallen end of the spring, and wedge it, by baulks upon the running board, in nearly its old position. With the help of ropes the thin angle-irons of the running boards might carry the additional weight.

We had on each car a length of scantling to place between the double tyres if ever the car stuck in sand or mud. Three blocks of this would make the needful height. We had no saw but drove bullets through it crosswise till we could snap it off. The Turks heard us firing and halted cautiously. Joyce heard us and ran back to help. Into his car we piled our load, jacked up the spring and the chassis, lashed in the wooden baulks, let her down on them (they bore splendidly), cranked up, and drove off. Rolls eased her to walking speed at every stone and ditch, while we, prisoners and all, ran beside with cries of encouragement, clearing the track.

In camp we stitched the blocks with captured telegraph wire, and bound them together and to the chassis, and the spring to the chassis; till it looked as strong as possible, and we put back the load. So enduring was the running board that we did the ordinary work with the car for the next three weeks and took her so into Damascus at the end. Great was Rolls, and great was Royce! They were worth hundreds of men to us in these deserts.[3]

The ability to remove and refit the square-headed tapered bolts holding the spring bracket would have made further repair possible.

Lawrence became a national hero due to the sensational reporting of the Arab Revolt by the American journalist Lowell Thomas. Returning to England a full colonel, Lawrence attended the Paris Peace Conference between January and May as a member of Faisal's delegation, and he was tipped for a knighthood. However, he was torn between a desire for fame and disgust at the politicians. Even Churchill noted his ambiguous tendency of 'backing into the limelight'. Later he became an object of satire: 'Clad in the magnificent white silk robes of an Arab prince … he hoped to pass unnoticed through London. Alas he was mistaken.'[4]

Eventually, sickened by the betrayal of the Arabs, Lawrence turned his back on the world and tried to enlist into the RAF as an ordinary aircraftsman under a false name. He was interviewed by the recruiting officer W. E. Johns (later known as the author of the *Biggles* series of novels). Johns rejected Lawrence's application as he suspected that 'Ross' was an alias. Lawrence admitted that he had presented false documents but returned later with an RAF messenger who carried a written order that Johns must accept Lawrence into the RAF. And so began a somewhat happier time for him when he helped develop high-speed air–sea rescue launches, dismantling engines and rebuilding them. In a letter to the poet and author Robert Graves he explained:

I went into the R.A.F. to serve a mechanical purpose, not as leader but as a cog of the machine. The keyword, I think, is machine. I have been mechanical since, and a good mechanic, for my self-training to become an artist has greatly widened my field of view. I leave it to others to say whether I chose well or not: one of the benefits of being part of the machine is that one learns that one doesn't matter!

One thing more. You remember my writing to you when I first went into the R.A.F. that it was the nearest modern equivalent of going into a monastery in the Middle Ages. That was right in more than one sense. Being a mechanic cuts one off from all real communication with women. There are no women in the machines, in any machine. No woman, I believe, can understand a mechanic's happiness in serving his bits and pieces.

Riding fast motorbikes assuaged his depression. Teddy Roosevelt also had suffered from the condition and had ridden his horses hard: 'Black care rarely sits behind a rider whose pace is fast enough.'

What was the key to this enigmatic man? Lawrence was an assumed name; he was the illegitimate son of an illegitimate daughter, and his father was an Irish baronet called Sir Thomas Chapman, who had eloped with his children's Scottish governess, taking on the name Lawrence. His mother believed in the purifying powers of corporal punishment and zealously beat her second son. These beatings had a deep psychological effect on Lawrence's character. He suffered from 'the conceit of self-abasement' and later used to pay a man to flog him. His brother maintained that this was his attempt to suppress rather than stimulate his sexuality in the manner of the ascetics of the medieval Church; he was obsessed by Thomas Malory's *Le Morte d'Arthur* as a boy. The oppressive family atmosphere of pretence and subterfuge may have facilitated his adoption of aliases throughout his life. He was a consummate exaggerator and embroiderer, but his mental distress after the war was real enough:

> This sort of thing must be madness, and sometimes I wonder how far mad I am, and if a mad-house would not be my next (and merciful) stage … When my mood gets too hot and I find myself wandering beyond control I pull out my motor-bike and hurl it top-speed through these unfit roads for hour after hour. My nerves are jaded and gone near dead, so that nothing less than hours of voluntary danger will prick them into life: and the 'life' they reach then is a melancholy joy at risking something worth exactly 2/9 a day.

Cresting a brow at high speed on his Brough Superior motorbike, Lawrence swerved to avoid two boys on their bicycles and was thrown off. He was not wearing a helmet and died, still in a coma, six days later, on 19 May 1935. His mourners included Winston Churchill, General Wavell, E. M. Forster and Siegfried Sassoon.

Henry Royce met Lawrence when, as Aircraftsman Shaw, Lawrence was working at the air base at Calshot where the Rolls-Royce-engined Schneider Trophy seaplanes were being prepared. Royce was staggered by the encounter. 'Who would have thought', he said, 'that little fellow was Lawrence of Arabia?'

His epitaph may have been coined for him by Aubrey Herbert, surveying his fellow members of the intelligence cadre brought together in Cairo in 1914: 'Then there's Lawrence, an odd gnome, half cad – with a touch of genius.'

T. E. Lawrence arrives in Damascus in his Rolls-Royce tender in 1919, driven by Sam Rolls.

CHAPTER NINE

Racing improves the breed

After the Armistice was signed on 11 November 1918 the aviation industry in Britain found itself with a sudden shortfall in orders, even though there was a growing popular enthusiasm for powered flight. Surplus aircraft could be bought cheaply, and barnstorming air shows for eager crowds provided employment for wartime pilots. Some of the larger aircraft began to be used for mail deliveries, and an air-travel industry was born for wealthy passengers. Rolls-Royce returned to motor-car manufacture and started to deal with a backlog of orders for the 40/50 Silver Ghost.

Then the 1920s roared in. 'It was an age of miracles,' F. Scott Fitzgerald wrote of the Roaring Twenties, 'it was an age of art, it was an age of excess ...'[1] Post-war euphoria combined with female liberation combined with a dizzying speed of change led to a glamorous decade for some – 'the whole upper tenth of a nation living with the insouciance of grand dukes and the casualness of chorus girls'. In Britain their older brothers might have been slaughtered in the war, but the new generation had a devil-may-care attitude. In Fitzgerald's words they were the beautiful and the damned. Jazz

A Vickers Vimy, powered by Rolls-Royce Eagles, shortly before taking off on the first successful flight from England to Australia in 1919.

defined the age with improvised rhythms and unruly spontaneity. Art Deco was the look, taking cues from the bold shapes of Cubism, the primary colours of Fauvism and the exoticism of Egyptomania: King Tutankhamun's tomb was excavated in 1922. Egyptian styles became popular for women's fashion, too: they flung out their Edwardian mothers' corsets, bustles and dresses and ditched their complicated long hairstyles and hats. What they wanted now was short bobbed hairstyles, silk pyjamas and short drop-waisted dresses.

For the vast majority of the British population who couldn't afford to live like grand dukes or buy Rolls-Royces, the most significant post-war social change was votes for women. Women's participation in the munitions factories and across the whole war effort finally dispelled the notion that they were physically and mentally inferior to men and gave the lie to claims that they were temperamentally unfit to vote. Age-old male anxieties about women in the public arena could no longer be justified.

Even though the Isle of Man had allowed some women with property to vote since 1881, it wasn't until February 1918 in Great Britain and Northern Ireland that women aged 30 were allowed to vote. In 1928 full parity with men was established, the franchise being then extended to all women over the age of 21. Elsewhere the Twenties brought more cars, telephones, cinemas, wireless and electricity. The popular new newspapers started to concentrate on celebrities such as film stars, sports heroes and aviators. And above all was the worship of speed. Ocean liners, planes, trains and automobiles all competed for Fastest and Furthest. And that brings us to the rise of the air race.

The well-connected Claude Johnson at Rolls-Royce was friendly with Alfred Harmsworth, Lord Northcliffe, the proprietor of the *Daily Mail*, the *Daily Mirror* and *The Times* newspapers. He was also an enthusiastic owner of a Rolls-Royce 40/50 Silver Ghost. The first of the buccaneer press barons, Northcliffe had made a fortune out of cheap newspapers for the masses before wireless, TV and the internet. Prime Minister Robert Cecil, Lord Salisbury, regarded the *Daily Mail* as dangerous populist propaganda, 'written by office boys for office boys'. Northcliffe was poorly educated, ignorant of history and science, but he had a lust for money and power. 'Every extension of the franchise', he had written in 1903, 'renders more powerful the newspaper and less powerful the politician.' He claimed to have won the First World War by overthrowing the Prime Minister, Asquith, and replacing him with Lloyd George. His intensely anti-German rhetoric led the *Star* newspaper to claim: 'Next to the Kaiser, Lord Northcliffe has done more than any living man to bring about the war.' So angry were the Germans that they sent a warship to shell his house at Broadstairs in an attempt to assassinate him – a large hammer wielded to crack a small nut. Instead the attack killed the gardener's wife.

Back in 1906 Northcliffe, sensing a public appetite for aviation, had offered a prize for the first long-distance air race, a flight from London to Manchester. The prize was £10,000, a vast sum of money worth around £1.2 million today. *Punch* magazine thought the whole idea ridiculous and also offered £10,000 for the first flight to Mars. However, the prize was won in 1910 by a Frenchman, Louis Paulhan, flying a Farman III biplane propelled by one of the new Gnome rotary engines that were to prove so successful in the First World War. The *Daily Mail* followed this up with a £1,000 prize for the first flight across the English Channel, won by another Frenchman, Louis Blériot. Then, just before the war, £10,000 was offered for the first non-stop flight across the Atlantic. This was interrupted by the conflict, but Northcliffe repeated the offer in 1919. Ten thousand pounds was still worth around £500,000 today after four years of wartime inflation.

There was huge national interest in the transatlantic attempt, and seven contenders started to prepare their aircraft: a Boulton Paul Atlantic, Sopwith Atlantic, Martinsyde Raymor, Handley Page V/1500, Short Shamrock, Fairey Special and Vickers Vimy. All of these were powered by Rolls-Royce engines, except for the Boulton Page, which featured the superb new W12 Napier Lion.

This aero engine had been designed by the Peterborough-born engineer Arthur Rowledge and built largely out of aluminium. Unusually, it had three banks of four cylinders spaced at intervals of 60° in a broad arrow, making it gloriously compact, light and rigid. It also had four valves per cylinder and double overhead camshafts, too, like the best cars of today. Too late for the war, it powered many of the aircraft and world speed record-holders of the interwar period. It was also more powerful than the Rolls-Royce engines, at 450 hp.

Jack Alcock, the pilot chosen for the attempt in the Vickers Vimy, made sure to visit Derby, where he spoke at length with Claude Johnson about the best speed to run the Eagle VIII. It turned out that the most effective revs were 1,650 rpm, the same speed as the first major critical torsional harmonic of the crankshaft. Thanks to Henry Royce's invention of the crankshaft damper, Alcock was able to maintain this rate most of the way across. He also recruited a quietly competent navigator, Arthur Brown, and discovered that they had something in common: they had both been shot down and imprisoned during the recent conflict.

In May 1919 the competing aircraft assembled on various fields in Newfoundland, hoping to ride the prevailing winds across the Atlantic. On 18 May the Martinsyde and Sopwith were ready for their attempt. It was a miserably cold day and the Martinsyde crashed within 200 yards of take-off. Harry Hawker, the pilot of the Sopwith (who would later give his name to the company that built the Hurricane), took off and flew into the cloud. During the night the single Rolls-Royce Eagle engine started to overheat and cut out. Hawker tried everything to cure the problem, including diving the Sopwith and stopping and starting the engine. The problem grew worse, so he turned south to seek shipping lanes and ditched near an eastbound steamer, the *Mary*. Thanks to the Sopwith's built-in lifeboat, Hawker and his navigator Kenneth Grieve were rescued.

The *Mary* had no radio, so the whole nation was on tenterhooks for days. The Sopwith seemed to have disappeared into thin air. King George sent telegrams of condolence to the families. Then a light signal was received at the Butt of Lewis: 'Saved hands Sopwith aeroplane.' The receiving station signalled: 'Is it Hawker?' The *Mary* replied: 'Yes.' The *Daily Mail* awarded Hawker and his navigator Grieve a consolation prize of £5,000, and on their return to London for decoration by the King they were met by excited crowds. They arrived at the palace in an open-top Rolls-Royce Silver Ghost. Hawker named his second daughter Mary in tribute to the ship that saved his life.

There remains a mystery: the failure of the Eagle engine. The floating wreck of the Sopwith was recovered by the US steamer *Lake Charleville* and found to have the radiator shutters firmly closed, which would have caused overheating. Hawker refuted this, saying he deliberately closed the shutters to avoid his navigator being scalded by boiling seawater when ditching. Also, if they had been closed the overheating would have happened soon after the full-load, full-throttle take-off, not halfway across 'the pond'. This overheating may in fact have been caused by ice blocking the radiator. The cutting-out and recovery of the engine could have been carburettor icing, a problem little understood then, and also experienced by Alcock and Brown on their flight. The courage of attempting a pioneer flight across an ocean with just one engine goes without saying.*

The Handley Page team undertook test after test while the Vickers team quickly assembled their aircraft and made ready. Beating the others, Alcock and Brown took off on 14 June, with the heavily fuel-laden Vickers Vimy staggering over the end of the field and just missing the trees. It was a tricky flight. First their air-driven generator failed, so they lost their electrically heated suits, their wireless and their intercom (curiously, the Eagle engines were not fitted with an electrical generator). Then an exhaust manifold fell off, causing such an unholy din that the two men couldn't hear each other in the open cockpit. Then they flew into thick fog, making it impossible for Brown to get a star sight.

Later they were forced to fly into thick cumulo-nimbus clouds, and turbulent winds flung the Vimy around. They were getting drenched by hail which was turning to ice on the wings and hence the aeroplane was getting heavier and heavier and nearly unflyable. At this point a fuel-flow gauge started icing up and Brown had to stand up out of the cockpit to clear it in the freezing gale. Suddenly the Vimy stalled and spiralled downwards through the storm: 4,000 … 3,000 … 2,000 … 1,000 feet … the aircraft continued its plummet through the cloud. Five hundred … 250 … 100 feet … They burst out of the clouds at only 60 feet to find the sea not beneath them but standing up on its side: they were falling sideways through the air.

---

* Having said that, spare a thought for the female bar-tailed godwit named E7 which was recorded flying across the Pacific Ocean from Alaska to New Zealand, a journey of 7,145 miles (11,500 km) in a single flight.

Instinctively Alcock centralised the joystick and rudder. The Vimy responded at once. Alcock opened the throttles wide, the Rolls-Royce engines roared and they regained their flying speed, skimming the surface of the waves, at times so close that the spray of the white horses beat on the underside of the wings. The danger was past.[2]

Except that the danger wasn't past. As they had no idea whether a faster plane was behind them, Alcock decided to land in Ireland instead of continuing to Brooklands, which would have been within fuel range. Mistaking an inviting-looking bog for grass, he put the aircraft down and crash-landed the Vimy, which promptly fell over onto its nose.

At last the Atlantic had been crossed. Later Alcock recalled a salty taste in his mouth that was foam from the sea …

After a journey like that it is easy to see why transatlantic flights didn't immediately catch on. No one repeated the feat until Charles Lindbergh in 1927. Meanwhile Alcock and Brown were lionised, knighted, and presented with the £10,000 prize by Winston Churchill. Decently they insisted that the Vickers and the Rolls-Royce mechanics who had helped them should receive a £2,000 share of it. Captain Alcock said in a speech to the Aero Cub: 'All the credit is due to the machine and particularly to the engines – that is everything.'

He was dead within six months, crashing a Vickers Viking flying boat in fog on a flight to France.

The accident rate of early aviation makes for shocking reading and is a reminder how far we've come. With impressive foresight the British Air Ministry decided in 1918 to encourage long-range flights to stimulate the fledgling civil aviation industry. The successful Alcock and Brown transatlantic flight was followed by another competitive flight from London to Australia. The Australian government offered £A10,000 for the first Australian pilot to fly the route, but the competition was marred by the usual crop of crashes and fatalities.

An Alliance P.2 Seabird took off from Hounslow Heath on 13 November 1919, but it failed to gain enough altitude and crashed a few miles away in Surbiton, killing both pilots. A Martinsyde Type A disappeared over Corfu and only one of the pilots' bodies was found. The Sopwith Wallaby and the Blackburn Kangaroo both crashed on landing, but a Vickers Vimy made the whole distance, landed at Darwin and claimed the prize. It was another success for the Rolls-Royce Eagle engines* which ran for 135 hours without overhaul, most unusual then (now over 2,000 hours is commonplace for a modern aircraft piston engine).

Now it was time to open up an air route from London to Cape Town. The Air Ministry had spent a year preparing no fewer than 24 aerodromes and 19 emergency landing strips. Several expeditions declared their intention to fly the route, with a Vickers Vimy sponsored by *The Times* leaving first on 24 January 1920. They had endless trouble with their Eagle engines, suffering water leaks and overheating. Finally, the starboard engine failed on take-off from Tabora and the Vimy swerved off and destroyed itself in the bush. Apparently some regrettable language was used. The *Daily Telegraph* had sponsored a Handley Page which smashed its tail and three wings, and a DH14 was written off on its second forced landing on 24 July 1920. However, the crew of a second Vickers Vimy named *Silver Queen I* showed incredible persistence. This one had been sponsored by the government of South Africa, as General Smuts wanted South Africans to be the first to fly the route. He authorised the purchase of a Vickers Vimy, G-UABA, for

---

* Numbers 5466 and 5716.

£4,500 (about £200,000 today). It took off from Brooklands on 4 February and the crew had a traumatic 11-hour non-stop crossing of the Mediterranean – the first – which took place in terrible weather. The pilot, Lieutenant Colonel van Ryneveld, described it as 'an unforgettable nightmare … an ugly impression which we would like to obliterate from our minds'.

Things got worse. During a night flight from Heliopolis the drain cock on the radiator of the starboard engine vibrated to the open position, allowing all the cooling water to escape and the engine to overheat (one wonders why the cock wasn't wired closed). They had to make a forced landing in pitch darkness north of Wadi Halfa. On landing, the aircraft hit a pile of boulders and was wrecked, but the crew escaped serious injury. Undeterred, the Rolls-Royce mechanic on board, F. W. Sherratt, removed both Eagle engines and carted them back to Cairo, whereupon they were fitted into a second Vickers Vimy that was perhaps unwisely named *Silver Queen II*. This one got to Wadi Halfa, where the fuel tanks were accidentally filled with water, meaning that they had to drain the whole system. The language employed has not been recorded. Unbowed, they took off again, but engine troubles delayed them at Shirati on the eastern shore of Lake Victoria, then again at Livingstone. Then on take-off at Bulawayo the Vickers Vimy, overloaded with fuel, failed to gain altitude and crashed, being totally wrecked this time. The two pilots and two mechanics somehow escaped serious injury. General Smuts telegraphed that another aeroplane was being sent and van Ryneveld and Quintin Brand took off in a DH9, which they flew to Cape Town without incident. They landed on 20 March 1920, the first men to fly from England to Cape Town, for which achievement both of them, like Alcock and Brown, were knighted.

As a postscript to this appalling epic, a pilot who regularly flies from London to Lusaka remarked to the author that the main problem with the route nowadays is coping with the boredom.

CHAPTER TEN

Air racing's golden age

The Schneider Trophy races of the 1920s provided a turning point in Rolls-Royce's progress as an aero-engine manufacturer and helped with the design of the future Merlin engine. Although the company's engines had built up a fine reputation during the war and, as we have seen, powered the great pioneer flights across the Atlantic and from Britain to Australia and Cape Town, by the 1920s Rolls-Royce was concentrating on motor cars and had slipped into third place behind Napier, with its brilliant W12 Lion engine, and Bristol, with the mighty air-cooled radial Jupiter. But then thanks to the intense development of the Rolls-Royce R racing engine for the Schneider Trophy the tables had turned, and by the end of the Twenties the company had reasserted its place as the makers of the best aero engines in the world. Furthermore, the work done by Reginald Mitchell in designing his Schneider Trophy S5 and S6 Rolls-Royce-engined seaplanes led directly to his design for the Spitfire and his choice of the Merlin to power it.

The Supermarine S6.B won the Schneider Trophy in 1931 and set a new world speed record shortly afterwards. Reminiscent of Turner's *Rain, steam, and speed.*

Mitchell had served a five-year apprenticeship on a steam-loco-motive firm until landing a job at the Supermarine Aviation Works in Southampton. This company, the future builder of the Spitfire, had been founded by Noel Pemberton Billing, who chose the name Supermarine as he wanted to concentrate on aircraft that would fly over the sea as opposed to submarines that operated beneath it. Pemberton Billing was one of those extraordinary individuals that are hard to believe really existed. Running away from his Hampstead home at the age of 13, he ended up in South Africa, where he became in turn a mounted policeman, a boxer, an actor, and was present at the Relief of Ladysmith. Returning to England, he passed his exams to be a lawyer, then instead decided to become an aviator, betting Frederick Handley Page £500 that he could earn his pilot's licence within 24 hours of first sitting in an aeroplane. This bet he won, allowing him to set up as an aircraft manufacturer just before the First World War. His first effort, the PB-1 Supermarine flying boat and the direct ancestor of the Spitfire, was fitted with a feeble 50 hp Gnome rotary engine. It floated but refused to take off. Only one was built.

Pemberton Billing then sold the company and became a bellig-erent Member of Parliament, offering a boxing duel to a fellow MP and agitating for a separate Air Force independent of the Navy or Army. In 1916 he invented the fake news that the Germans were blackmailing '47,000 highly placed British perverts' listed in an imagined Black Book to 'propagate evils which all decent men thought had perished in Sodom and Lesbia'. The Germans were also targeting wives of politicians – 'in Lesbian ecstasy the most sacred secrets of the state were threatened'. He hinted that Lady Asquith, the wife of the Prime Minister, was a lesbian. Minds boggled. Billing then published an article, 'The Cult of the Clitoris', which implied that the actress Maud Allan was a lesbian associate of the conspirators. Allan had performed Oscar Wilde's *Salome*,

'Bare-limbed and scantily draped in filmy gauzes', gurgled the *New York Times*. 'Miss Allan ... is more beautiful in face and figure than most ...' The article led to a libel case, at which Billing represented himself and, astonishingly, won.

The new owner of Supermarine, Hubert Scott-Paine, was a more sober individual, at first concentrating on the manufacture of aircraft under licence. Then, after the war and wanting to build his own designs, he promoted Mitchell at the age of only 24 to the post of Chief Designer. The rest of the Twenties saw Mitchell concentrating on building winning designs for the Schneider Trophy. Scott-Paine later worked with T. E. Shaw (aka Lawrence of Arabia) to develop the RAF high-speed 'crash' boats which during the Second World War would rescue some 13,000 lives. Lawrence showed a rare empathy with the boats' Meadows engines, machines that would neither worship him nor betray him.

The Schneider Trophy is an important part of our Merlin story because the 'R' racing engine built by Henry Royce and his team for the 1929 and 1931 competitions introduced advances in mechanical construction, supercharging, and fuels that were to stand Rolls-Royce in good stead ten years later. As Arthur Sidgreaves, the managing director of the company, commented later, 'research for the Schneider Trophy contest over the past two years is what our aero-engine department would otherwise have taken six to ten years to learn.' It also became a prestigious international showcase for the aviation industry.

Jacques Schneider was a keen balloonist and aeroplane enthusiast. He was also the heir to an armaments empire which enabled him to afford his ballooning activities and to set a world altitude record at 33,074 feet (10,081 metres). Like Charles Rolls, he gave up ballooning when he saw Wilbur Wright demonstrating his aeroplane at Le Mans in August 1908 and took up the exciting new sport of aviation. A serious accident curtailed his own flying, but after this he supported many air races, the most memorable of which was his speed contest for seaplanes, the 'Coupe d'Aviation Maritime Jacques Schneider'. Schneider believed that seaplanes were the future, as stretches of water such as lakes and bays provided ready-made runways. Competitors had to fly at least 150 miles (240 km) over open sea on a triangular course, and the prize for the victor was 25,000 gold francs (about £1,000) and a cup worth the same as the prize. If a nation won three times in a row the trophy would belong to them for ever.

The Schneider Trophy races were truly international, as opposed to the national races such as the Royal Aero Club's in Great Britain and the Pulitzer races in the USA, and the leading industrial nations competed intensely to produce the most slippery airframes and the most powerful engines. More often than not it was a poor airframe that lost a race, but it was invariably the best engine that won it. This resulted in a worldwide acceleration in the progress of aero-engine design, as you may see by looking at the winning speeds over the years until 1931. By then the speed had climbed to 340 mph (550 km/h), an increase of 7½ times in around 20 years. This reflected the enormous pace of progress in the aero industry.

The first Schneider Trophy was held on 16 April 1913, at Monaco, and it was won by a French Gnome-engined Deperdussin monoplane at an average speed of only 46 mph (74 km/h), partly because some of the distance was covered taxiing on the water when the pilot landed too soon! This was an extraordinarily prescient design using a stressed-skin construction (of plywood) and a monoplane configuration that would prove to be the way to go in the future. The next year the British won with a Sopwith Schneider biplane at nearly double the speed, at 87 mph (140 km/h). There was a shambolic contest at Bournemouth in heavy fog in 1919 when everyone got lost, then the Italians – the only contenders – won at Venice in 1920 when the speed was 107 mph. They won again in 1921.

1922 was a more convincing contest for three reasons: (1) if the Italians won again they would keep the trophy for ever, (2) all the entrants finished the course, and (3) the average speeds were much higher than previously, showing serious intent on the part of the competitors.

Supermarine entered their amphibious biplane Sea Lion II, which was about as aerodynamic as Buckingham Palace, but because it was fitted with the 450 hp Napier Lion it just scraped a win over the Bay of Naples at 146 mph. The contest could continue. The following year, in 1923 Sea Lion II came third to the US entry: the pace was hotting up. The Americans had come up with an excellent new design that was going to prove to be influential to the engines that powered both Allied and Axis sides of the next war: the Curtiss D-12 engine. Instead of separate cylinders with leaky and expensive separate water jackets, each line of six cylinders was cast into an aluminium block, an idea first tried on the Hispano-Suiza V8. This made the engine immensely rigid, light and leak-free. The two cylinder blocks were set on a common crankcase in a V-shape at 60°, making it a V12 configuration. However, it didn't have reduction gears and it was never effectively supercharged. It looked rather like the future Merlin.

The story of the Curtiss D-12 aero engine in Britain illustrates the vicious competition between British aero manufacturers for the few scraps of work thrown to them by the Air Ministry. When the American engine appeared at the 1923 Schneider Trophy competition it had impressed the British aero-engine and aeroplane builders. One in particular, the aircraft manufacturer Richard Fairey, was noted for building elegant sharp-nosed aeroplanes such as the Fairey III, which he had sold in hundreds. He preferred

water-cooled engines as they gave a clean-looking nose and
reduced drag compared with big radial engines, and he had been
using the Napier Lion to good effect. He went to the United States
to have a closer look at the Curtiss D-12, and what he found was a
compact all-aluminium V12 of around 19 litres with a small frontal area that would suit his designs perfectly.*

He was so impressed that he bought a licence to build the
Curtiss D-12 and returned to England determined not only to
make the engine but also to design a new bomber around it as a
private venture funded by the profits from his Fairey III. The result
was the Fairey Fox, which at once got Air Ministry backs up as it
was faster than their newest and latest fighters. Furthermore, it had
not been designed to an official specification, and they objected to
details such as its internal fuel tanks and the fact it had an American
engine. However, on seeing the prototype Fox being demonstrated,
Air Chief Marshal Hugh Trenchard, the Chief of the Air Staff,
turned to Fairey and remarked: 'Mr Fairey, I have decided to order
a squadron of these machines!', thus bypassing official channels.
An order for 18 Foxes swiftly followed to equip 12 Squadron. Such
was the performance of the Fairey Fox with the Curtiss engine that
the squadron was instructed to fly no faster than 140 mph (225
km/h) during annual Air Defence Exercises in order to give the
defending fighters a chance. It seems that somewhere, though,
Fairey had made an enemy.

There were no further orders for the Fairey Fox from the RAF,
and when the Air Ministry issued a new Specification 12/26 for a
high-performance bomber which the Fox would suit, they
neglected to send a copy to Fairey. He only received the specifica-

---

* Rover historians will remember the all-alloy American Buick V8 being
licensed for use in Rover saloons, Land Rovers and Range Rovers in a similar
way.

tion after protesting to the Air Ministry. What he wanted was the government to back him in producing the Curtiss D-12, but the government wasn't having it. They had four engine manufacturers on their books and they wanted no more. They liked the Fairey Fox but they wanted it fitted with a British engine built by one of the established companies. Bristol were busy making their excellent air-cooled radial, and Armstrong Siddeley were making their smaller air-cooled engines. Napier didn't want to do anything other than keep selling their Napier Lion, and so it was Rolls-Royce, by a process of elimination, that the Air Ministry approached. Sir Geoffrey Salmond, air member for Supply and Organisation, sent up a Curtiss D-12 engine and a contract and asked them to give them something like it, but better. And so it was that Henry Royce and his team started the design of the Falcon X, which became the F and eventually the Kestrel. And it was that engine, enlarged 20 per cent, that eventually became the Rolls-Royce Merlin.

The Rolls-Royce test pilot Ronald Harker saw the Kestrel in the test cells:

Sometime during my apprenticeship in 1926 an unfamiliar exhaust noise began to come continuously from the aero engine test beds. It was quite unlike the normal muted boom coming from the slow-revving Eagles and Condors, which had their exhausts silenced. This new sound was much more like a racing car, higher revving and with open stub exhausts. It came from the experimental test bed, which was out of bounds to the general workers. By peeping through a crack in the fence, I was able to see what this new engine looked like, while it was running on the test stand. It was a V twelve-cylinder engine, all aluminium and elegant, and much smaller than the other aero engines. It looked to be a great improvement on the older engines.[1]

This 21-litre Kestrel was developed by Arthur Rowledge, the man who had designed the Napier Lion and who had left Napier after a disagreement. He took the best of the Curtiss: the V12 configuration, monobloc aluminium cylinder blocks, four valve heads and supercharging at all altitudes. Less desirably it also had direct drive, but Rowledge soon fixed this, adding reduction gears that allowed the engine to rev higher. The test pilot Harker observed:

Its only weak point being a propensity for internal water leaks; these occurred at the top cylinder joint between the cylinder liner joint and the aluminium cylinder block. It was a weak feature in the design which persisted right into the Merlin and was only really overcome when a two-piece cylinder was developed having a dry joint between the cylinder head and the block. Fortunately, the effect of an internal coolant leak was not damaging to the engine nor dangerous to the pilot or aircraft.[2]

Under full throttle with high cylinder pressures the water would be vaporised and passed out of the exhausts, but when the pilot throttled back the coolant seeped in faster and could short out the sparking plugs, causing misfiring. On landing the cylinder block would have to be removed – a huge job for the fitters.

Apart from this failing, the Kestrel engine became a great success, giving 520 hp by May 1928 and powering the Hawker Hart biplanes that dominated British air power in the early 1930s.

Fairey didn't do well out of all this manoeuvring. The Air Ministry had made assurances that it intended to use the Rolls-Royce Kestrel in further Fox aeroplanes, but it seems these were intended to encourage Rolls-Royce to get on making the Kestrel with their own money while the government tried to find some more from the taxpayer. The first batch of Kestrels were fitted to two more squadrons of Fairey Fox and the following year the aircraft was dropped and replaced by the Hawker Hart, to which subsequent models of Kestrel were devoted. The Rolls-Royce Kestrel went on to well-deserved fame, but it should be acknowledged that once again the company had copied a good design ... and then made it better.

Aeroplane airframes were steadily growing more beautiful as they became more streamlined and simplified. Antoine de Saint-Exupéry was a French aristocrat, an existentialist, an aviator and a lyrical writer on the subject of flight:

Have you looked at a modern airplane? Have you followed from year to year the evolution of its lines? Have you ever thought, not only about the airplane but about whatever man builds, that all of man's industrial efforts, all his computations and calculations, all the nights spent over working draughts and blueprints, invariably culminate in the production of a thing whose sole and guiding principle is the ultimate principle of simplicity?

It is as if there were a natural law which ordained that to achieve this end, to refine the curve of a piece of furniture, or a ship's keel, or the fuselage of an airplane, until gradually it partakes of the elementary purity of the curve of a human breast or shoulder, there must be the experimentation of several generations of craftsmen.[3]

Back at the Supermarine drawing office Reginald Mitchell realised that only a revolutionary airframe design would have any hope of competing in the next Schneider Trophy race. The draggy old biplanes made of canvas, struts and wire were nearly coming to the end as racers. Wing loading is the weight of an aircraft divided by the wing area, and the large area of the twin wings made the wing loading low, giving good manoeuvrability and a low stall speed: good characteristics for a fighter plane. But the huge air resistance presented by the twin wings of the biplanes made them too slow. The fact that the Schneider racers also had to drag along two ridiculous great floats to land on just meant that the designers had to try harder. So for the 1925 Schneider Trophy Mitchell drew the Supermarine S4, a revolutionary clean monoplane design that was so expensive to build that the Air Ministry had to subsidise the costs. It was powered by a 680 hp Napier Lion VII engine.

It is worth mentioning at this point that the Gnome and Bentley rotary engines that were so successful in the First World War had also reached the end of their development. Having a heavy spinning engine at the front of the aeroplane multiplied gyroscopic precession. More powerful rotaries were even heavier. This resulted in control and stability problems, especially for inexperienced pilots. The way to go seemed to be with V-shaped water-cooled engines with a small frontal area, allowing a clean-nosed airframe. V12-powered monoplanes were the future.

The Schneider Trophy was now becoming something of an international power race, with Mussolini pouring money into the Italian entries. However, the Supermarine S4 crashed in Baltimore bay in 1925, with the pilot Henri Biard reporting that the wings had suffered from high-speed buffeting. This was a new phenomenon for Mitchell to deal with. The Americans won with their Curtiss R3C-2, the last gasp of the Schneider biplanes, propelled by a Curtiss DV-12. Again, the race had been won by the best aero engine.

Then the Italians won again in 1926 with their V12-engined monoplane Macchi M.39. This was another monoplane with design features tailored for the Schneider Trophy competition. The course circuit required left turns, so the left wingtip was slightly farther from the fuselage than the right wingtip to allow it to make tighter left-hand turns. When taking off, the opposite and equal reaction to the twist of the propeller was nearly submerging one float of the prototypes, so to counteract this propeller torque reaction the floats had unequal buoyancy.

Lovers of Venetian gondolas may know that they too are asymmetrical and tend to steer to the left. They are deliberately built with a 24 cm wider left side of the hull to compensate for the weight and propulsion from the gondolier, who stands on the right.

There was no entry from Britain in 1926, but Mitchell came back in 1927 with an improved design: the S5, a Napier Lion-powered monoplane. It won in Venice. Everyone took a deep breath and agreed on a two-year delay before the next Schneider Trophy. The British and the Italians were going to be fighting tooth and nail for victory.

Mitchell knew that although he now had a superior airframe he now needed a lot more power for the next contest in 1929 – twice the power. Deciding that the Napier Lion engine had reached the end of its development, he sat down with Rolls-Royce's new managing director and Major George Bulman from the Air Ministry: please would they build a racing engine for the next Schneider Trophy? In those days Rolls-Royce still had a dislike of racing. If they entered the Best Car in the World in a race and it lost, what would that make the winner? There were still unpleasant memories of the embarrassing Austrian Alpine Trial. They held the same view on aero-engine racing, and thank you, they didn't want the job. Oh, that Milton should have been there, wrote Setright, quoting the poet: 'I cannot praise a fugitive and cloistered virtue, unexercised and unbreathed, that never sallies out and sees her adversary, but slinks out of the race, where that immortal garland is to be run for, not without dust and heat.'[4]

On hearing that Rolls-Royce didn't want the commission, the persuasive Major Bulman retorted: 'Bollocks. I order your firm to take on this job.' And so it was that Rolls-Royce got the contract and built the R-type engine that contributed so much to the Merlin.

When told of the commission Henry Royce appeared confident that they could pull it off and promised Mitchell at least 1,800 hp. Royce was an invalid by then and three of his colleagues went down to see him at his home and headquarters at West Wittering in October 1928. They found him enthusiastic:

> It was a bright autumn morning and Royce suggested a stroll along the beach; as they walked he pointed out the local places of interest. But Royce, who walked with a stick, was a semi-invalid … and he soon tired. 'Let's find a sheltered spot,' he said, 'and have a talk.'
>
> Seated on the sand dunes against a groyne, Royce sketched the rough outline of a racing engine in the sand with his stick. Each man was asked his opinion in turn, the sand was raked over and adjustments made. The key to the engine was simplicity. 'I invent nothing,' was Royce's philosophy, 'inventors go broke.' … The secret of increased power would lie in supercharging.[5]

So just how did Rolls-Royce progress from the 360 hp Eagle to the 1,900 hp R engine?

'The 1929 Rolls-Royce R typifies Rolls-Royce engineering. There is nothing much about the design that was original or even especially meritorious, but the development work was absolutely marvellous.'[6]

At the end of the First World War Rolls-Royce had produced a galumphing great 35-litre monster of an engine, the Condor, which was essentially a scaled-up Eagle intended for bombers.

They hadn't done much with it, being too busy selling 40/50 Silver Ghosts and leaving the field wide open for the Napier Lion and Bristol radial engines. The Condor had four valves in each cylinder and fork and blade connecting rods, both features that eventually went into the Merlin. The four valves allowed more air-and-fuel mixture into the cylinders and more exhaust gases out, which allowed extra power to be developed. The connecting rods between pistons and crankshaft in an engine are usually the most stressed component and cause most catastrophic failures. When a rod breaks at high revolutions it usually goes out through the side of the cylinder block, smashing a hole and often starting a fire. 'Throwing a rod' or 'putting a leg out of bed' caused the deaths of many aircrew.

V-engined car engines usually place the rods from each bank of cylinders side to side on the crankshaft-bearing journal, but this results in narrower, stressed big-end bearings, longer, more bendy crankshafts and a nasty 'rocking-couple' vibration caused by the two banks of cylinders being staggered fore and aft. The more expensive Rolls-Royce fork and blade solution harked back to marine steam-engine practice and was, typically of Rolls-Royce, a Proper Job. However, the otherwise old-fashioned Condor was never a success.

But then, lo! Condor begat a son in his own likeness: the 37-litre Buzzard engine, which featured monobloc cylinder blocks similar to the Kestrel and the American Curtiss D-12 engine, plus the four-valve heads and fork and blade connecting rods. It was running in June 1928, producing around 800 hp, and it was in fact the Buzzard that became the sire of Schneider Trophy-winning R engine, as it was the most suitable big engine they had to hand. It is worth pointing out that this line of ancestry is distinct from the line that resulted in the Merlin, which was descended directly from the Kestrel.

Henry Royce was confident in his deep-breathing valves and the strong bottom end that descended from the Condor, so to these went into the sketch in the sand at West Wittering beach. But the device that doubled the power of the Buzzard and made the R racing engine a winner was sketched in at the back of the engine: a supercharger.

If God had intended humanity to fly he might have been more helpful and made air density the same at all altitudes. Climbers on the summit of Mount Everest at 29,028 feet (8,850 metres) are sucking in air at about a third of the pressure at sea level, and each breath therefore contains only a third of the oxygen. As a result they go blue in the face and fall over unless they get a reviving draught of bottled oxygen. Aircraft engines feel much the same, losing power as they climb. Aircraft cannot carry enough bottled oxygen, but they *can* force extra air into the engine using an air pump called a centrifugal supercharger, or blower. This is a disc with vanes on it which can be driven directly from the crankshaft (or by a turbine in the stream of exhaust gases, in which case it is called a turbocharger). The R supercharger was driven eight times faster than the crankshaft and sent a whirling gale of wind into the cylinders. More oxygen was added to the chemical reaction of combustion, and thus more power was developed at the propeller. Rolls-Royce experimented with a double-sided supercharger that would just fit inside Mitchell's new Supermarine S6 airframe, and in evolving the R engine they also added a Kestrel-style air intake at the front of the aeroplane which would ram air in and increase the pressure further. This alone accounted for 10 per cent of extra boost, which at 350 mph (563 km/h) was equivalent to as much as 250 hp.

* * *

Conditions in the Rolls-Royce engine-test cell in Derby were horrific. As well as the R engine running at full throttle there were three Kestrel engines also running flat out, one to simulate a 400 mph headwind and providing the ram effect, one cooling the crankcase, and one outside in the yard driving a propeller to blow away the fumes. The racket was heard as far away as Nottingham and the people of Derby started to protest at the 24 hours a day testing. The mayor of Derby had to step in and ask the populace to put up with it for the sake of British prestige. The eight men in the test cell were given cotton wool to protect their ears, but they suffered from deafness and tinnitus. The engines were using vast amounts of castor oil, and this was being ejected from the exhaust ports and dripping down the walls; the men were given milk in an attempt to reduce the laxative effects. But now the Rolls-Royce testers ran into yet another problem: detonation. And to understand that we need to look at the fuel they were using.

As a teenager I used to play with a neighbour's single-cylinder diesel engine, the sort of thing that was used to drive a circular saw. It was about the size of a horse and when hand-starting it had the kick of a mule. It was dangerous, but I learned interesting things about internal combustion. The engine would run on all sorts of unexpected fuels: if I dripped cooking oil into the inlet port and turned off the fuel pump it would keep running. I read that the Shetland Bus fishing boats would run on seal oil, but lacking a seal (this was inland Rutland) I tried creosote, butane gas, paraffin and petrol. This last produced a death-rattle which was alarming: detonation. You may sometimes have heard this as a petrol-fuelled car engine passes in the street in too high a gear: a sharp metallic rattle. Some call it pinking, or knocking. Henry Royce, the former scarecrow, called it chirruping. Rolls-Royce engineers could detect the onset of detonation by holding a short piece of steel between their teeth and resting the other end on the cylinder block of an engine running at full throttle.

Combustion in a cylinder is usually initiated by the sparking plug at a specific point in the piston's travel; between 10 and 40 crankshaft degrees before the piston comes to the top of its stroke (Top Dead Centre). The flame-front should spread smoothly, burning all the fuel/air mixture just in time to push the piston back down the cylinder. But detonation happens when pockets of fuel/air mixture explode before they should, as Royce realised in 1913. The shockwave causes the characteristic pinging sound as combustion pressures rise sharply. The results can be disastrous: first overheating and then the erosion of holes in pistons or cylinder heads, and this can be fatal to an aircraft and its crew.

Henry Royce's first cars would have run on a solvent that was commonly available at chemists' shops. The classically trained British called it petroleum, which comes from the Latin *petra*, rock, *oleum*, oil: literally rock-oil. This was appropriate because petroleum was found under sedimentary rocks where countless bodies of zooplankton and algae had been subjected to enormous pressure and heat. It was fossilised fuel (and the energy was fossilised sunlight). The Chinese had a refinery during the Song dynasty, which they named the 'Fierce Oil Workshop', but the first modern oil well wasn't in Texas or Saudi Arabia, but at the Riddings colliery at Alfreton, Derbyshire. In 1847 a chemist, James 'Paraffin' Young noticed a seepage of oil and managed to distil from it a light oil that he could use in lamps instead of expensive whale oil. His patented discovery probably saved whales from total extinction.

Different fractions of the distilled product were useful for different things: paraffin for lamps and the more volatile petrol as a solvent or fuel. The Americans called the fuel gasoline, or confusingly 'gas' for short, a name that derives from the British coffee merchant and Temperance publisher John Cassell. Seeing new opportunities for artificial light, in 1862 he set up an oil refinery in Hanwell, London, and marketed his new lamp fuel under his own name: 'The Patent Cazeline Oil, safe, economical, and brilliant … possesses all the requisites which have so long been desired as a means of powerful artificial light.'[7]

Sales boomed, then fell away in Ireland. Cassell found out that a Dublin shopkeeper, one Samuel Boyd, was selling counterfeit Cazeline Oil and asked him to stop. The shopkeeper responded with a paintbrush, changing every label to read 'Gazeline', coining a word that eventually spread throughout the USA. Cassell took him to court and won, but Boyd's coinage won in the end: 'gasoline'.

When used in Rolls-Royce 40/50 engines, ordinary chemist-shop petroleum worked well enough until the engine's compression ratio was increased beyond around 3.2:1. This ratio is a measure of how tightly the fuel and air is squeezed in the cylinder. More accurately, it is the ratio between the volume of the cylinder plus the combustion chamber when the piston is at the bottom of its stroke, and the volume of the combustion chamber when the piston is at the top of its stroke. So if a Silver Ghost cylinder drew in a deep breath of 1 litre of air/fuel mixture it could safely compress it to only a third of the volume. Any higher a ratio, and detonation would set in: Henry Royce's bird-like 'chirruping'. Modern car engines use a compression ratio of around 10:1 because there are huge gains in power and thermal efficiency to be had. The Silver Ghost made around 48 horsepower from over 7 litres, a pitiful 6.8 hp per litre, whereas a modern General Motors LS 7-litre engine of the same capacity, with an 11:1 compression ratio, makes 505 horsepower; 72 hp per litre, or more than ten times as much. A lot of the difference has to do with the advances in fuel because they enabled higher compressions to be used.

* * *

Fuel quality was of vital importance to Rolls-Royce and other aero-engine manufacturers, and it is vital to our story. A Liberty engine of the First World War had the same capacity of 27 litres and the same V12 configuration as a Merlin engine of the Second World War. However, the former developed 400 horsepower and the latter eventually as much as 2,340 horsepower. The difference would not have been so great if fuels hadn't improved enormously during the intervening years: we can attribute no less than half the power gains to improved fuel chemistry, bearing in mind that the other improvements such as supercharging could not have been employed with poor-quality fuel. By stuffing twice the air/fuel volume into a cylinder using a supercharger the effective compression ratio was doubled, which with poor fuel would lead to appalling pinking, chirruping, knocking, detonation, and then disaster.

How did they improve the fuel? Petrol's resistance to detonation is measured in octane number: the higher the octane number the more compression the fuel can withstand before detonating. The 40/50 Silver Ghost of 1910 would have been running on petrol of about 40 octane, and during the First World War aircraft engines would have used petrol of around 50 octane. Petrol bought in a supermarket petrol station today might have an octane rating of 95 or so. If the octane number is increased from 72 to 100, the power is likely to increase 100 per cent. The converse is true; when tests were done in 1937, an octane reduction of 13 points (from 100 down to 87 octane) decreased engine performance by 20 per cent and increased take-off distance by 45 per cent: potentially fatal for aircraft and crew.

Low-octane fuel nearly brought disaster to the Allies during the First World War. Before the American entry into the war their European allies used petrol distilled from crude oils from the Far East which gave acceptable performance in their aircraft engines.

When the United States entered the war in April 1917 the US became responsible for supplying petrol to the Allies, and suddenly a decrease in engine performance was noticed. Engines gummed up and sparking plugs fouled. If full throttle was used, detonation set in, and a number of aircraft were lost. Panicked messages were sent across the Atlantic, and it was found that petroleum from aromatic and naphthenic base crude oils were superior. These came from California, South Texas and Venezuela.

Exotic fuels were cooked up for the Schneider Trophy Rolls-Royce R engines by Rodney Banks, a British fuel chemist working for the Anglo-American Oil Company. The winning engine of 1929 ran on 78 per cent benzol, 22 per cent Romanian petrol with a dash of tetraethyl lead, and it developed 1,900 hp with high supercharger pressures. In 1931, when the trophy was won for perpetuity, the fuel chemists concocted a brew for a world speed record attempt consisting of 60 per cent methanol, 30 per cent benzol and 10 per cent acetone, and the Rolls-Royce testers were able to screw up the supercharger pressure to make a staggering 2,530 hp.

These exotic brews were impracticable for military use, being expensive and unstable, so petroleum had to be made with higher octane rating, and this is where Thomas Midgley Jr comes into our story. He was an American engineer working for General Motors who discovered in 1921 that adding the substance called tetraethyl lead (TEL) to petrol prevented detonation or knocking in engines. His team had an engine on the test-rig running under detonating conditions when suddenly: 'The ear-splitting knock of their test engine turned to a smooth purr when only a small amount of the compound was added to the fuel supply. and all the men danced a non-scientific jig around the laboratory.'[8]

The problem was that this TEL compound was highly poisonous, and they knew it. All mention of lead was omitted from publicity materials, but Midgley fell ill from lead poisoning. He had to drop all work in 1923 and take a long vacation in Miami. General Motors created the Ethyl Corporation and built a plant in New Jersey to mass-produce the stuff. Workers suffered hallucinations, insanity, and five deaths. To prove that his compound was safe, Midgley participated in a press conference in 1924 at which he poured TEL over his hands, placed a bottle of the chemical under his nose, and inhaled its vapour for 60 seconds, declaring that he could do this every day without succumbing to any problems. Shortly afterwards he had to take absence from work after again being diagnosed with lead poisoning. The State of New Jersey ordered the plant to be closed a few days later.

Still, the profit motive prevailed, and what little regulation there was in the US seemed ineffective. TEL was manufactured and spread worldwide as a constituent of high-octane petrol, releasing large amounts of lead into the atmosphere and causing countless cases of brain damage, particularly among children. Violence and criminality rose in inner cities and IQs fell, partly due to lead.

Not content with that, Midgley was on the General Motors team that invented chlorofluorocarbons (CFCs) for air conditioners and refrigerators under the name of Freon, which was later implicated in the destruction of the ozone layer of the atmosphere. Once again Midgley flamboyantly demonstrated the safety of his lethal brew, this time in front of the American Chemical Society, by inhaling a breath of the gas and using it to extinguish a candle. Midgley was highly decorated for his work before it was discovered to be so diabolical, and he was duly elected the president and chairman – of the American Chemical Society.

Midgley has been described as the single organism that has had the most negative impact on the world's atmosphere, ever, and Bill Bryson wrote in 2003 that he possessed 'an instinct for the regrettable that was almost uncanny'.[9]

Poor Thomas Midgley Jr came to a sad end. He was diagnosed with polio in 1940, which confined him to bed. Inventive to the end, he devised a system of pulleys and ropes to manoeuvre himself, and in 1944 became entangled in this and died of strangulation.

Environmental protests prevailed, and after lead in petrol was banned a 2011 study by the California State University found that 'ridding the world of leaded petrol ... has resulted in $2.4 trillion in annual benefits, 1.2 million fewer premature deaths, higher overall intelligence and 58 million fewer crimes'. Once again, thanks to scientists and experts who spotted the ozone hole caused by CFCs, another Midgley menace was eradicated. In 2007, 200 countries agreed to eliminate hydrochlorofluorocarbons entirely by 2020. Finally, the dead hand of Thomas Midgley Jr had been lifted.

Back at the 1929 Schneider Trophy, the days on the banks of the Solent before the contest proved to be even more nerve-racking for the Rolls-Royce team than the race itself. Engine problems cropped up, and so parts and whole engines were driven down from Derby at high speed overnight in an adapted Rolls-Royce Phantom pick-up truck known as the 'Phantom of The Night'. On the evening before the race, while checking over one of the S6 aeroplanes, Cyril Lovesey's attention was drawn to a tiny blob of silvery aluminium on one of the plugs. This indicated that a piston was on the verge of seizing. Rod Banks was there:

A pre-race inspection of the newly-installed race engines revealed some aluminium on the plugs removed from a cylinder on Waghorn's machine. This indicated that the particular piston was probably picking up. According to the Schneider Trophy Contest rules, once the race engines were finally installed they could not be removed; but the Royal Aero Club Contest Committee members at Calshot agreed that components could be removed and replaced in situ in the aircraft.

Wisely, Hives had brought a coachload of Rolls-Royce engineers down from Derby, and they were rounded up from hotels and pubs. The whole aircraft had to be tilted 30° in order for the hoists to lift off the cylinder block and damaged piston:

> They worked through the night, removed the cylinder block, changed the offending piston, replaced the block and had the engine running at 8 a.m. Waghorn, who won the Contest, was not told until afterwards about the night's happenings.[10]

After more hair-raising moments the race was won by the same repaired aircraft at a speed of 328.63 mph (529 km/h) over the course.

Somewhere in the RAF contingent at Calshot that day was Aircraftsman Shaw, the man once known as Lawrence of Arabia. He had first met his Commanding Officer, Sydney Smith, during the Cairo Conference, and the two had become friends. Lawrence's duties included working as Smith's personal assistant during the organisation of the 1929 Schneider Trophy seaplane contest. One wonders how he felt as the winning Supermarine S6 thundered over the line.

Rolls-Royce were so delighted with the result that they presented Mitchell with a Rolls-Royce car. Britain had now won the contest twice in a row. One more victory in 1931 would clinch it, and the Schneider Trophy would be theirs in perpetuity. Henry Royce was created a baronet in 1930 for his services to British aviation. He chose the name of Seaton in the county of Rutland. Seaton was where his family had once been millers. (I know the village well;

the old railway station is home to a scrapyard where I spent many happy hours dismantling engines.)

But then, on Tuesday 29 October 1929, the bottom of the world fell out. The Wall Street Crash was heard all around the world. It plunged Britain into the Great Depression and it signalled the end of the Roaring Twenties. Days before, the economist Irving Fisher had reassured the public that stock prices had reached 'a permanently high plateau'. Scott Fitzgerald commemorated the decade: 'The ten-year period that, as if reluctant to die outmoded in its bed, leaped to a spectacular death in October 1929'.[11]

Reginald Mitchell and Henry Royce at Calshot. Mitchell looks every inch the thrusting young designer with slicked-back hair; Royce is taller, older, bearded and more authoritative. They were not to know that they shared something other than that 1929 victory; they were both to do their best work after being diagnosed with bowel cancer and after undergoing a colostomy.

CHAPTER ELEVEN

*The female of the species is more
deadly than the male[1]*

After her third husband died mysteriously, Lady Lucy Houston found herself extremely wealthy. She stood at the rail of her steam yacht *Liberty* moored in the harbour at Cowes on that day in 1931 and gazed at the tiny speck racing across the Solent towards her. Standing next to her was her favourite stepson with two of his chums. 'Bring two of your friends, but they must be good-looking,' she had stipulated. The speck resolved into a racing aeroplane, which now banked hard left and raced in the opposite direction. Thanks to Lady Houston it was now a winning aeroplane; the Supermarine S6B was to be developed into the Spitfire and its engine helped the birth of the Rolls-Royce Merlin.

In the late 1920s the Royal Air Force was fighting for its life. After the horrific First World War and under pressure from those advocating peace, the British government was disarming. The country's aircraft industry was fragmented and in ferocious competition. At the last minute, and in response to the economic depression of 1929, the government refused to support the 1931 Schneider Trophy air race. It was only because of a generous gift

Lady Lucy Houston.

from Lady Houston* that Supermarine could afford to enter the competition and win the trophy for perpetuity. Two years afterwards she sponsored the first-ever flights over Mount Everest, and later her understanding of the vital importance of air power led her to make an offer of funding for several squadrons of aircraft for the defence of London. This offer was rejected by the government.

It is easy to make a case for this fervently patriotic woman as being the single most important individual in the winning of the Battle of Britain. This is another of the Rolls-Royce legends. But is it true?

Biographers often have an axe to grind; either a shiny chopper that reveals how marvellous their subject was, or less often a small sharp hatchet with which to dismember a reputation. Lady Lucy Houston had such an extraordinary life that either interpretation would be possible. On the face of it she was a beautiful woman who made the best of her gifts, a suffragette, a philanthropist and patriot. On the other hand she was a fortune hunter, a social climber and a fascist sympathiser.

Fanny Lucy Radmall was the seventh of ten children, two of whom died. Her father was a reasonably prosperous London woollen-draper who lived over the shop with her mother Maria, whose profession was 'wardrobe dealer': clothes rather than their receptacles. They lived over the shop near St Paul's cathedral where Lucy played hide and seek among the gravestones. She was a wild child, and a pretty one too, 'a creature of tremendous vitality and utterly roguish charm, with tiny hands and feet, a wasp waist and large impish eyes'.[2]

* Apparently the surname Houston should be pronounced How-ston – not like the Texan mission control.

At the age of 16 she became a chorus girl known as Poppy, and worked the boards for only six weeks before attracting the attention of a 34-year-old toff, Frederick Gretton, who was the wealthy scion of the Burton brewing family. He whisked her off to Paris where they lived together as man and wife for ten years. They never married, even though they clearly loved one another, so it is possible that he was secretly married already. It was an exciting time to be in Paris in 1873 just after the Franco-Prussian War, with the Impressionists exhibiting for the first time and the salons filled with high society. Lucy, now known as Mrs Gretton, swiftly learned French and hobnobbed with the great and the good. And the less than good, too: she was introduced to Edward, the roué Prince of Wales, who would often be seen at the theatre with his latest mistress, one of at least 55 conjectured liaisons. He was on the whole a humane and charming man, and although they didn't become intimates, Lucy's meeting with 'Bertie' made a deep impression upon her and she was an admirer of the monarchy from then on.

Her lover Frederick Gretton became one of the most successful racehorse owners in Britain, but then at only 43 he suffered a stroke, lingered a few months, then died. He left Lucy an annuity of £6,000 (around £780,000 today). She was 26, beautiful and wealthy.

Lucy's love of social status may have informed her next choice of partner, Lieutenant-Colonel Sir Theodore Francis Brinckman, 3rd Baronet, whom she divorced in 1895. Working herself steadily through the pages of Debrett's she next married an impoverished peer of the realm, George Frederick William Byron, 9th Baron Byron. Forty-five-year-old 'Red-nose' George had drunk and gambled away what fortune he had, and Lucy's money was very welcome to him. She in return became Lady Byron. During their marriage she became an active suffragette, and shortly after he died in 1917 she was appointed Dame Commander, Order of the British Empire (DBE) for her support of a home for nurses who had served in the First World War. She was deeply disturbed by the news from Russia that the royal family had been deposed by the Bolsheviks. This turned to horror at the news of their murder, and it made her vehemently anti-Communist for the rest of her life.

In 1920 the Labour Party announced a general strike if the British government supported the White Russians' civil war against the Bolsheviks. The suffragette movement opposed this strike and a newspaper reported that 'Lady Byron is lending the full weight of her influence … and has called on all women to refuse to cook, clean and cater for the strikers.' She was quoted as saying that 'the time has come for women of all classes to unite solidly and to demand that the Government brings in a law making all strikes illegal'. Certainly the strike had threatened to bring the country to its knees, and Lucy decided she needed more money for her next political excursion. Although she was still wealthy from Frederick Gretton's will, as she turned 64 she was on the prowl for another husband.

Sir Robert Houston, 1st Baronet, Conservative Member of Parliament for West Toxteth, was an immensely wealthy and somewhat unlikeable man. He is described in the *Oxford Dictionary of National Biography* as 'a hard, ruthless, unpleasant bachelor'. He had made his fortune running a fleet of ramshackle ships down to South Africa during the Boer War, but he was also a connoisseur of fine yachts. His latest and finest was the *Liberty*, a luxury steam yacht of 93 metres and 1,450 tonnes. She had been built in Leith in 1908 for Joseph Pulitzer, the newspaper magnate, who like Northcliffe had made a fortune out of cheap 'Yellow' journalism. Pulitzer's depression, blindness and acute sensitivity to noise led to his retirement onto the *Liberty*, whose cabins were therefore heavily sound-insulated and whose furniture featured rounded edges. He died on board in 1911 aged only 64, muttering in German: '*Leise, ganz leise*': 'Quiet, very quiet.'

Houston bought *Liberty* in 1919 and spent three seasons cruising with his friends Sir Thomas Lipton of the tea company and Sir Thomas Dewar of the whisky brand. One evening whilst on board Lipton's yacht *Erin*, the two elderly bachelors hatched a plot against their friend. Terrified of Lucy, whom they regarded as a loose cannon, they decided to marry her off to 'Black Bob' Houston. They arranged an introduction in 1922, and Lucy latched onto him at once: 'I determined to marry him the moment I set eyes on him; he was a real man: never wasted time or money.' Over the next two years she was a regular guest on *Liberty* on cruises to Monte Carlo and Cannes in company with the great and the good, such as the Russian dukes and duchesses who had been deposed by the Bolsheviks. She met Grand Duke Cyril, who would have been the next Tsar of Russia, and from whom she heard the horrific stories of butchery, torture and murder that disfigured the Bolshevik revolution. She became more and more convinced that her personal mission was to try to stem the tide of Communism that was threatening to spread across Europe into her homeland. She deeply admired Mussolini and wondered whether a man like him might be found to bring his ideas to the British Empire.

The past is a foreign country: they do things differently there. It might be difficult for us now to understand the appeal of fascism to British people such as Lady Lucy Houston. We look back through a murky past filled with Benito Mussolini, Adolf Hitler, General Franco, the Nazi Party, the Gestapo and the horrors of the Holocaust. But as Boris Johnson says of the period: 'In the 1930s your average toff was much more fearful of bolshevism, and communists' alarming ideology of redistribution, than they were fearful of Hitler.'[3]

Fascism might have seemed particularly appealing to the aristocrats and officer classes who had seen their sons slaughtered in the trenches, fighting a country they saw as a natural ally, over someone else's problem in the Balkans. Wasn't the Kaiser Queen Victoria's grandson? They deplored the moral vacuity of the Bright Young Things of the new generation and the weakness of the elected Members of Parliament. Those who had ruled in India were used to autocratic rule instead of democratic rule, when things could get done quickly and efficiently without reference to Whitehall. Just like the Princely States, where Maharajas ruled by feudal authority, the British Viceroy exercised unlimited power over his subjects. One such was the Conservative Lord Curzon, who had reigned as Viceroy of India from 1899 to 1905. On his return to England he had served in the War Cabinet and saw the dangers of universal suffrage to his party. He became the leader of the Anti-Suffrage League, resisting the Votes for Women movement, but he was out of step with the times.

In 1920 his daughter Cynthia married a Tory MP named Sir Oswald Mosley, who became the nearest the British have had to a fascist dictator. He crossed the floor and became an independent, then joined the Labour Party and was part of their first government in 1924. In their second term of government he became frustrated by a lack of political power, and so he resigned and formed his New Party, which inclined more and more to the right and adopted strongly anti-Semitic views in imitation of Hitler's policies. He created the British Union of Fascists in 1932, which received approval and support from Viscount Rothermere, who in January 1934 wrote – under his own byline – articles that appeared in both the *Mail* and the *Mirror*. The former was headlined 'Hurrah for the Blackshirts', whilst the latter had 'Give the Blackshirts a helping hand'. Both newspapers planned a beauty contest aimed at finding Britain's prettiest woman fascist, an idea that might not fly today. Rothermere admired both Hitler and Mussolini, meeting Hitler and congratulating him on his annexation of Czechoslovakia.

'Depend upon it, sir,' Dr Johnson said to his biographer Boswell, 'patriotism is the last refuge of the scoundrel.' By this Johnson meant the self-professed patriot, the rousers of popular nationalism such as Hitler and Mosley, and the scoundrels seen more recently in modern politics. We can thus distinguish between the patriotism of Oswald Mosley and Lady Lucy Houston's love of her country.

Oswald Mosley used to affect a Nazi-style military uniform, he was a great orator, and he had many affairs with women. During his marriage he began an affair with his wife's younger sister Lady Alexandra Metcalfe and with their stepmother, Grace Curzon. Lucy was also bowled over, thinking this young politician who was adopting her hero Mussolini's policies might be her 'man of destiny'. However, Mosley had misjudged the tolerance of the British. In October 1936 he attempted to march his Blackshirts through an area of London with a high proportion of Jewish residents, and violence flared up between protesters trying to block the march and police trying to force it through. This was the Battle of Cable Street, and for Lucy and many others it was end of her dalliance with Oswald Mosley.

Back on *Liberty* 'Black Bob' was having sinking feelings about Lucy. 'Mark my words, Goodie,' Houston growled to his skipper, Captain Goodwin, 'she's after my bloody money.'[4]

For her birthday in 1924 he had sent a tray of jewels for her to select a present, but she ticked him off for sending trinkets of too little value. She went to Bond Street and instead chose a string of black pearls worth £50,000 (£3 million today). Houston had met his match in Lucy.

Houston's health was declining, and he had the first of several strokes in 1924. Lucy promised to supervise his nursing and it was possibly this that finally broke down his resistance. They married in Paris in December 1924 and he bought a house in Jersey to avoid the tax regime in England. While taking the spa waters at Harrogate, Houston suffered another stroke, and to get him back to Jersey quickly and to avoid the possibility of his dying in England, Lucy chartered a train to Southampton and then a 1,000-passenger ferry – just for the two of them. Once there, Houston wrote a draft will and showed it to Lucy, remarking that he had left her a million pounds. Lucy tore it up and threw it on the floor. 'If I am only worth a million pounds, then you had better go away and remake your will and leave it to someone who deserves it better than I do!'[5] Houston duly rewrote the will and ultimately left her £4½ million (£270 million).

Houston had made many enemies in his lifetime, and Lucy became convinced that someone was trying to poison him. When he did die in 1926 she was so prostrated with grief that she was unable to attend the funeral. There was a vast sum of money at play in his estate, and she was declared insane by the Jersey Royal Court, which concluded that she was 'suffering from delusions of persecution'. They effectively put her under house arrest, appointed a curator to manage her affairs, and when *Liberty* arrived at Jersey, impounded the yacht. However, Lucy produced her own doctors from England, who declared her sane, and she eventually escaped Jersey on *Liberty*. She left a large Celtic cross memorial to Black Bob, which reads: 'This cross is erected by his sorrowing wife to the memory of Robert Paterson Houston, Baronet, who died most mysteriously on the 14th of April 1926. "My Robert, my dear, dear Robert."'

'Most mysteriously'? The way is clear for a modern-day biographer or detective to fathom out what exactly happened to Black

Bob and his fortune. *Cui bono*? – who stood to benefit from his death? Lucy was now stupendously rich, and able to indulge ambitions that were both philanthropic and highly political. The British government felt that Houston's estate ought to be liable to English death duties, and Lucy had a running battle over this for the next few years. She loathed the new Labour government of 1929, and their most reprehensible action in her eyes was a further reduction in funding for the Royal Air Force.

After Britain's victory in the Supermarine Trophy of 1929, Prime Minister Ramsay MacDonald had stood up at the celebrations and promised financial support. Then after the Great Depression set in a few months later he withdrew his promise. The government would not contribute one penny to the costs of the 1931 Schneider competition. This might now seem a reasonable response to a doubling of unemployment, a halving of exports, widespread poverty and hunger marches. Aviation races might have appeared to be a luxury for the few.

Lord Rothermere's *Daily Mail* group of newspapers launched a public appeal for money, and several thousand pounds were raised. Once again, they were supporting advances in aviation. Lucy sensed an opportunity to both embarrass her enemies and support her country and its aviation industry. She wrote a cheque for £100,000 (over £6 million today) and presented it to the Air Ministry. This was received with ill grace, but the cheque was cashed. Rolls-Royce could now press on with the development of the R engine, and Supermarine with a modified S6, the S6B, for the 1931 Schneider Trophy.

When British government support for the Supermarine was withdrawn, the Italians were delighted. Benito Mussolini personally supported the Italian Schneider Trophy efforts in an attempt to gain international prestige for his nation's aircraft industry. He poured money in at the same time as the British government withdrew its support. Macchi Aeronautica came up with the Macchi MC72, which proved to be the fastest seaplane ever built. It was powered by a stupendous Fiat engine of no less than 24 cylinders and 51 litres. In fact, this was two V12 engines arranged in tandem on a shared alloy crankcase, which became a structural member in the airframe. This configuration ensured a small frontal area. The two crankshafts were kept separate to avoid torsional vibration problems and each drove a concentric shaft in the V of the cylinder blocks forward to the twin counter-rotating propellers. This was a clever solution to the torque-reaction problems found with powerful single-propeller machines. Engineer Mario Castoldi had studied contra-rotating propellers for years, learning that the efficiency of the rear airscrew was not impaired by the front one; on the contrary it was up to 20 per cent more efficient. The surplus power of the rear engine could thus be used to drive the supercharger for both V12s.* The Italians often built beautiful engines and the power output of this beauty was 2,800 hp in 1933 and eventually 3,100 hp in 1934. The fuselage seemed almost an afterthought, and the whole effect was that of an enormously big engine hotly pursued by a tiny wisp of an aeroplane.

Italian V12 engines always seemed musical. The writer was working on a 1970s Ferrari V12 when he noticed an extra exhaust tube exiting after the confluence of the individual exhaust pipes from the exhaust ports. It rejoined the system in the silencer. It was the same on the other side. Curious, I blanked it off and discovered that the engine had lost its tuneful howl. Further research suggested that Ferrari had added this extra pipe on each side of the engine to

---

* It is worth pointing out that, like Rolls-Royce, the Italians had bought a Curtiss DV-12 engine to examine.

enhance the odd harmonics in the exhaust sound. As any classical musician knows, a clarinet is a cylindrical-bore instrument closed at one end by the vibrating mouthpiece (and in the car the exhaust valve provides the source of vibration). The normal resonant modes have a pressure maximum at the closed end (the mouthpiece) and a pressure minimum near the first open key or the bell. These conditions yield odd harmonics in the clarinet's sound. A saxophone with its conical tube provides even harmonics, which is why the saxophone has a bright sound, and a clarinet produces sounds that are warm and dark. This is maybe one reason why a good piston engine can arouse musical feelings.

Supermarine and Rolls-Royce had little time to prepare for the 1931 contest, as they were given Lucy's green light in January for a race in September. The airframe could only be tweaked, and so it was down to Rolls-Royce to find more power to beat the Italians. This was done by gearing up the speed of the supercharger and opening up the ram air intake until the boost was 18 lb/in above atmospheric pressure. Sodium-cooled exhaust valves were used, molten sodium inside the hollow stem transferring heat from the head to the stem by splash (this idea was invented by the British engineer Sam Heron after reading that sodium had been used in thermometers). Meanwhile the fuel chemists concocted a brew of 70 per cent benzol, 20 per cent Californian petrol and 10 per cent methanol plus a few cc of lead. Even details such as the moisture content of this brew were carefully checked and rechecked. Eventually the engine would run for an hour in the test cell giving 2,350 hp.

Getting rid of the heat proved more of a problem; the S6B effectively became a huge radiator, with the engine oil even being sprayed inside the tail fin for cooling and water-coolant pipes running beneath the metal skin of the fuselage. Pilots flew the S6B watching the water temperature gauge as the limit to power, not the revolutions counter.

In the event the Italians could not get their engine reliable in time: backfiring problems plagued the double V12 engine, and after two pilots were killed – first Giovanni Monti and Stanislao Bellini – their 1931 Schneider entry was withdrawn. Two years later, after help from the fuel chemist Rodney Banks, the MC72 broke the world speed record for piston-engined seaplanes, when Francesco Agello, the last surviving test pilot of four, piloted the aircraft to an average speed of 440.7 mph (709.2 km/h). That record for a piston-engined seaplane still stands.

It is a shame that the Italians were unable to compete. Their design virtuosity was possibly superior to that of the British, but their development skills were clearly behind Rolls-Royce's. To finish first, first you must finish, and no aero engineer exemplified that philosophy better than Henry Royce. The Fiat engines, as beautiful as they were temperamental, were characterful predecessors of the sweet little racing V12s built by Ferrari after the Second World War.

The French team also suffered the death of one of their test pilots and withdrew their entry. No other contenders appeared for the 1931 Schneider Trophy race, and so it was a walkover for the Supermarine S6B.

The British pilots could have cruised round and won, but the team competed sportingly, running the Rolls-Royce R engine at the maximum revs of 3,200 and only reducing revs when the water temperature started climbing. The pilot, Flight Lieutenant John Boothman, flew the plane watching the water-temperature gauge like a hawk: getting rid of excessive heat always proved to be a problem with this aircraft. But the S6B behaved perfectly, going around the circuit at Lee-on-Solent at 340.1 mph, and 38 minutes later winning the trophy for Britain for perpetuity.

Lady Lucy Houston with the Schneider Trophy pilots at Calsot in 1931.

Shortly afterwards another S6B piloted by Flight Lieutenant George Stainforth went out and captured the world speed record at 379.1 mph, and two weeks later, with the R engine giving the utmost, he raised this to 407.5 mph (655.67 km/h).

Lady Lucy Houston became a national heroine, but sailed away on *Liberty* to Paris to escape the floods of congratulations. Her part in the story was immortalised in the 1942 film *The First of the Few*, but few have heard of her today.

* * *

We are not yet finished with the indomitable Lucy. In 1932 she decided that Britain's air defences were being dangerously neglected, and so she wrote out a cheque for £200,000 (£13m) to pay for several squadrons of fighter planes to defend London. This sum would have paid for 70 aeroplanes. The Cabinet rejected the offer because of conditions Lucy had set on how the money was to be spent. She was furious and ordered *Liberty* to anchor off Portsmouth with a huge illuminated sign hanging in the rigging.

As darkness fell a message to the Prime Minister became apparent: *DOWN WITH THE TRAITOR MACDONALD!*

An infuriated harbour master arrived and demanded that either the message or the vessel be removed. *Liberty* then sailed to Southsea, where the message now read:

TO HELL WITH THE TRAITOR MACDONALD!

Lucy then sailed to Poole before the authorities could catch up with her.

If this early example of electronic trolling wasn't enough, Lucy then went about inventing the first pirate radio station: she made plans to install a powerful wireless station in *Liberty* and anchor her just outside the British three-mile territorial limit, broadcasting anti-government propaganda. She owned the *Saturday Review*, an anti-government newspaper expressing popular nationalistic views. As she aged she became even more eccentric, demanding that her chauffeur drive her white open Rolls-Royce flat out. When overtaken by a smaller car she would shout: 'Call yourself a chauffeur? Get a move on! You're crawling along as if you're going to a funeral!'[6] When she arrived at a public house she would order a pint of Guinness to be brought out to the car. This she would then drink while the surrounding street urchins clustered around and took in the sight. She can also be credited with being one of the inventors of the bikini: when staying at her home in Jersey she would lie on the beach wearing nothing but a bra and the briefest of shorts.

After the success of the 1931 Schneider Trophy in achieving high speed, the next challenge seemed to be that of achieving high altitude. Lucy's forward thinking about aviation now brought her into another arena: the Mount Everest high-altitude flight.

It had become evident during the First World War that aircraft that could fly higher than their opponents held a considerable advantage: fighters could dive out of the sun just as the German Fokkers had, and high-flying bombers could avoid anti-aircraft fire from the ground and outpace fighters climbing in pursuit.

The idea of flying over Mount Everest had first been suggested by that most remarkable high-altitude pioneer Alexander Kellas. He was a Scottish chemist, explorer and mountaineer who had spent more time over 23,000 feet (7,010 metres) than anyone else on earth. On his way to many Himalayan summits he had pioneered the employment of Sherpas at high altitude. The ancient Greeks knew that the body would deteriorate at high altitude, but no one knew why until the late nineteenth century, when it was realised that low levels of oxygen led to a condition known as hypoxia. The advances made during the First World War in aero-engine design meant that pilots struggled to stay conscious at the higher altitudes being achieved, and there were many losses of pilots and aircraft as a result of hypoxia. Kellas spent the war at the Air Ministry working with Professor J. B. S. Haldane on solutions to this problem: superchargers might work for aero engines, but pilots needed reliable bottled oxygen and heated clothing.

Kellas was also the author of an article in the *Geographical Journal* in 1918 on 'the Possibility of Aerial Reconnaissance in the Himalaya', in which he suggested that an aerial reconnaissance of

Mount Everest might be the key to a successful attempt. It is ironic, then, that it was he who was the first victim of the Mount Everest expedition of 1921, dying of dysentery, still a day's march from the mountain he always longed to see.[7]

The 1933 Houston-Mount Everest Expedition was dreamed up by Lieutenant-Colonel Latham Valentine Stewart Blacker, an officer with the Indian Army. He had worked for the Bristol aero-engine company and was aware of the high-altitude performance of their supercharged Pegasus radial engine. He recruited the adventure writer Colonel John Buchan and got the backing of the well-connected pilot Lord Clydesdale. They received the offer of aircraft from Westland and engines from Bristol, but funding was not forthcoming. The British economy was descending into depression, and flying over Mount Everest did not seem essential to the government.

Lord Clydesdale then traded on his mother's acquaintance with Lady Lucy Houston and persuaded her to pay for the expedition. 'It was the prospect of raising British prestige in India through the expedition that appealed to her immensely.'

In the event the flights over the summit of Mount Everest were successful, the only hiccough being the accidental severing of the cine-cameraman's oxygen pipe on the first flight. He missed his shots of the summit but, against Lucy's orders, a second flight was made and successful footage secured.

Lucy was delighted by the worldwide press attention – 'I want it to be thoroughly understood by everyone that our chief aim in this adventure was to show India that we are not the degenerate race that their leaders present Britain to be. India will now be forced to realise that the British lion is still full of pluck and courage and this conquest of Everest is a splendid achievement by we Britons, and the people of India can be justly proud of it.' It is unclear if the people of India were proud, but even the Pope sent Lucy his congratulations. Lucy was delighted. Once again she had succeeded in embarrassing the government.

Although furthering her political aims, the technical achievements of her expedition were more lasting. Oxygen-breathing apparatus, electrically heated clothing, reconnaissance photography and the Bristol Pegasus engines themselves had all been tested in rough field conditions, and vital experience was gathered that would be needed in the next war.

Lucy's death was as odd as her life. George V died in 1936 and his son Edward, a rather different character, acceded to the throne. His private secretary wrote that 'for some hereditary or physiological reason his normal mental development stopped dead when he reached adolescence'.[8] Although he demonstrated courage during his visits to the front line during the war, much of his behaviour seemed inappropriate for a future king. He had conducted an affair with the courtesan Marguerite Alibert, who was notorious for shooting her Egyptian husband at the Savoy Hotel (just after returning from a performance of the operetta *The Merry Widow*).[9] To his father's disgust he had many affairs with married women. Edward was also racially prejudiced, believing that whites were inherently superior. In 1920, on a visit to Australia, he wrote of indigenous Australians: 'they are the most revolting form of living creatures I've ever seen!! They are the lowest known form of human beings & are the nearest thing to monkeys.'[10] These were the people, we now understand, who lived more in balance with nature than any other human beings care to. During his short reign his disregard for court protocol concerned royal officials and the Prime Minister Stanley Baldwin. The rise of fascism had interested him, and he consorted with Hitler and Mussolini.

Lucy admired Edward enormously and hailed him in her many personal letters as 'our Man of Destiny', a reference to the dictator Napoleon. She urged him to become a 'benevolent despot' in the fashion of Mussolini and Hitler. However, Edward had become infatuated with the American divorcee Wallis Simpson and insisted on marrying her. It later transpired that Joachim von Ribbentrop, the Nazis' foreign minister, had been Simpson's lover when he was ambassador to Britain in 1936, and she was in secret communication with him. The establishment could not stomach the prospect of a divorced woman with living ex-husbands becoming queen consort, and they may have been suspicious of Simpson. Neville Chamberlain wrote in his diary that she was 'an entirely unscrupulous woman who is not in love with the King but is exploiting him for her own purposes. She has already ruined him in money and jewels …'[11]

Lucy hoped that Edward would go ahead with the marriage and so oblige the government to carry out their threat to resign. He could then appoint a new government from his supporters and thus seize control of the executive. In effect she was encouraging him to take power as dictator. In the event King Edward VIII was forced by the establishment to abdicate, but it was a close-run thing. Lucy took this hard, stopped eating and ordered all the windows in the house to be left open. She died on 29 December 1936 in her eightieth year.

So what did Lady Lucy Houston actually contribute to British aviation and in particular to the Rolls-Royce Merlin? She was an intensely patriotic woman who saw how vital British air power was to the security of her country. She foresaw the horrors of the bombing of London and did something about it. Through her sponsorship of the Schneider Trophy and the Mount Everest flights she encouraged the development of enormously powerful aero engines, high-speed airframes and high-altitude aviation, all soon to be crucial in the approaching war on Germany. Did she help fund the predecessors of the Merlin? Yes. Was she responsible for the Spitfire? No. Dr Gordon Mitchell, Mitchell's son, wrote this of Lucy's sponsorship of the Schneider Trophy:

> It has been suggested that without Lady Houston's generous gift which enabled the 1931 race to be held, there might not be a Spitfire produced in 1936. However, as the S6B was essentially only a modified version of the S6, the major part of the vital experience R.J. (Mitchell) gained from the Schneider races, later to be of such value in designing the Spitfire, was in fact obtained in 1927 and 1929 with the S5 and S6.[12]

Lady Lucy Houston was a visionary, as was Charles Rolls. Both of them in their different ways saw the future and helped towards the birth of the Rolls-Royce Merlin.

CHAPTER TWELVE

Si vis pacem, para bellum –
If you want peace, prepare for war

During the 1930s the Rolls-Royce R engine at one point simultaneously held the world air, water and land speed records, an achievement never equalled before or since by any other power unit. The news was relayed to Sir Henry Royce at his house at West Wittering on the south coast.

Sir Henry had just finished repairing his local vicar's lawnmower. Standing back, quietly satisfied with his work, he ventured a remark that was paraphrased by the vicar's son, Eric Gill: 'Whatever is rightly done, however humble, is noble.' Eric Gill obtained access to Royce's house and carved upon the stone mantelpiece the Latin translation: *Quidvis recte factum: quamvis humile praeclarum.*

So goes another Rolls-Royce legend.

The sculptor Eric Gill was actually in his mid-forties when he became friendly with Royce and carved the inscription. He was also one of Britain's greatest and most influential artists, giving us the **Gills Sans** font that you may still see in the crisp lettering of the

Sir Malcolm Campbell with his Rolls-Royce R-engined race car *Bluebird* in 1933.

Rolls-Royce, BBC and Land Rover logos. At the height of his powers, he was soon to undertake a commission for a group of sculptures, *Prospero* and *Ariel*, for the BBC's Broadcasting House in London, but his subsequent reputation suffered somewhat when a biographer revealed details of his sexual abuse of his sisters, daughters, and dog. He might perhaps be remembered more for his belief that humanity still needed God in a materialistic civilisation, and for the need for intellectual vigilance in an age of triviality.

Why the Latin? Royce almost certainly muttered something English along the lines of 'It doesn't matter how small it is, it ought to be a proper job,' a sentiment with which any decent engineer would agree. But, as you might retort, '*Quidquid latine dictum sit altum videtur*' – 'Anything said in Latin sounds profound' – and it would seem that Royce attached importance to the appearance of a classical education just as he attached a classical portico to the front of his cars.

Royce strove for perfection in everything he did, and this is certainly one of the secrets of his company's continued success. Draughtsmen were expected to draw their designs repeatedly, rubbing them out and perfecting the details:

Royce used to refer to his men as 'Knights of the Rubber'. The amount of work put into each design was far in excess of anything practised elsewhere, and the utmost patience had to be exercised by all the members of his team when trying to meet the high standard which Royce insisted upon. Royce used to say that all the problems should be solved on the drawing board and when the part was made and tested 'it should only be to prove that the design was right'.[1]

This culture of perfection and intense development had certainly paid off with the successes of the 40/50 Silver Ghost, the Eagle, the Kestrel and the Schneider Trophy R engine. A pair of R engines were used by Sir Henry Seagrave in his speedboat *Miss England II* to attempt the world water speed record on Lake Windermere on Friday 13 June 1930. On his third run the boat hit a floating branch and flipped over. Seagrave's engineer was killed, and he himself died of lung haemorrhages shortly after being told that he had set a new record of 98.76 mph (85.82 knots; 158.94 km/h).

One R engine was fitted in Sir Malcolm Campbell's *Bluebird* car which established a new land speed record of 272 mph on Daytona Beach in 1933, and two R engines were fitted in Captain George Eyston's *Thunderbolt*. Eyston remembered Rolls-Royce's covert help with his earlier Kestrel-engined *Speed of the Wind* with gratitude:

> Apart from owning Silver Ghosts in the early 1920s, my real contact with Derby (where the Rolls-Royce headquarters was located) lasted through the 1930s, luckily beginning with the acquaintance of Ernest Hives, who took a lively interest in the experimental cars. I went to see him one day, hoping to get hold of a Kestrel engine for my car, *Speed of the Wind*, with which I hoped to obtain some long-distance world's records in America. He was very interested and found me one which had been used on the test beds to blow fresh air over the Schneider Trophy R engine during its tests. The problem was: how to install an aero engine in a motor car? With a sump (crankcase) touching the ground! Of course, it meant redesigning the whole of the lower half crankcase [and oil pumps, incidentally] which I did, but then I had to have parts made. At last a friendly foreman took compassion; he said the job would be done, and if I came one night all the finished articles would be thrown over the factory wall![2]

This particular Kestrel engine had already done many hours in a flying boat before serving as a cooling fan for the Rolls-Royce R engine, so it was no Kestrel chick. Despite this it took the 12-hour and 24-hour records at Utah, raising the latter record to 140.52 mph (226.15 km/h). The foreman, Bob Coverley, rose to manager of Rotol, which made many propellers in the Second World War, so his unofficial approach to factory support did him no harm.

Captain Eyston's later *Thunderbolt* record-breaker had two Rolls-Royce R engines and weighed a monstrous seven tons. It took the land speed record at 357.2 mph at Utah in 1938. This was only 0.2 mph slower than the world air speed record set by Squadron Leader Augustus Orlebar using the R-engined Supermarine S6. The public perception of the Rolls-Royce R engine could surely rise no higher:

'Of the size only of an office desk ... this 12-cylinder super-charged racing engine is more powerful than an express locomotive. Its design is stated to be so valuable that it is still on the Government's secret list.'[3]

Henry Royce knew that hard times were coming for his car business, but because of the worldwide fame of the R engine he now had something to sell in the aero-engine business. He felt there could be a market for a larger version of the successful Kestrel engine that was giving good service in the 1928 Hawker Hart bomber and the 1931 Hawker Fury fighter. But these were fabric-covered biplanes festooned with struts and wires, and aerodynamically clean monoplanes like the Supermarine S6 were clearly coming. In October 1932 Royce authorised the development of a new engine larger than the Kestrel but smaller than the Buzzard and incorporating as many of the lessons of the Schneider Trophy-winning R engines as possible. There was no contract forthcoming from the Air Ministry, but Royce knew that a new engine had to be developed for the next generation of fighters. This was a brave decision, as it would have to be funded by Rolls-Royce themselves: a private venture. The code name was thus PV12. It was one of the last decisions Royce would make.

Germany was ready to listen to a fascist dictator in the 1930s, even if Britain was not. Defeated in 1918 but not occupied, a narrative of betrayal arose after the signing of the Treaty of Versailles. The consequences of this treaty are still argued over to this day. Suffice to say that one of its provisions required Germany 'to accept the responsibility of Germany and her allies for causing all the loss and damage'. This Article 231 became known as the War Guilt clause, and it justified the demand for enormous reparations totalling £6.6 billion (£284 billion in 2019). Germany also had to disarm and return territories.

There was unease in many quarters. Marshal Ferdinand Foch, Supreme Allied Commander, thought the Germans were being treated too leniently and, with disturbing precision, predicted that this was not a peace, it was merely an armistice for 20 years (mind you, he also said that aeroplanes were interesting toys with no military value). On the other hand, the British economist John Maynard Keynes described the treaty as a punitive Carthaginian peace and stated that the reparations were excessive. The compromises in the treaty satisfied no one.

Because Germany was never occupied, the populace failed to understand the scale of their defeat, and were thus able to blame others such as Communists and Jews for their later hardships. A stab-in-the-back myth arose, the *Dolchstoßlegende*, which has a particular resonance for Germans. In Richard Wagner's 1876 opera *Götterdämmerung* the hero Siegfried is murdered by his enemy Hagen by being stabbed in the back with a spear. The hated 'backstabbers' were later singled out for punishment under Hitler's rule. The inflation of 1923 brought great suffering: a loaf of bread that cost around 160 Marks at the end of 1922 cost 200,000,000,000 Marks less than a year later. After stabilisation of the currency, things took a turn for the worse during the Depression of the early 1930s.

Germany was now listening to Hitler. When he came to power in March 1933 the country began a covert build-up of her armed forces. Even though the Treaty of Versailles had forbidden an air force and navy, these strictures were now ignored. Pilots were sent to Russia and Italy for military training. 'We now had the opportunity of training our fighter pilots with the Italian Air Force,' stated Adolf Galland, one of the top German fighter pilot aces; '… in order to avoid international complications for Italy as well as for Germany, the whole affair had to be treated with the greatest possible secrecy and carried out under rigorous camouflage.'

A few aircraft had been developed ostensibly for civil use, such as the 1931 Heinkel He 51 biplane, which masqueraded as a trainer but was in fact a fighter. This machine was fitted with the first BMW V12 engine, the 600 hp V1.* Then in 1932 the Heinkel He 70 Blitz monoplane appeared, which pretended to be a mail plane but was in fact the direct ancestor of the Heinkel 111 fast bomber.

Hermann Göring, a First World War fighter ace with 22 victories, was appointed as the Luftwaffe's commander-in-chief, but he had three weaknesses: his arrogance, his poor understanding of aviation technology, and an addiction to morphine contracted

---

* The BMW V1 also propelled the bizarre Rail Zeppelin high-speed railcar. This extraordinary device was made of aluminium, was styled like an airship, featured an open propeller at the rear, and still holds the world speed record for the petrol-fuelled rail vehicle at 143 mph (230 km/h). The advisability of using an open propeller in crowded railway stations and the difficulty of reversing consigned the Rail Zeppelin to the history bin marked Interesting but Dangerous.

during treatment for injuries sustained during the Beer Hall Putsch. To inject this he used a specially made gold syringe.

As they were building their squadrons from scratch in 1933 the Luftwaffe planners were able to study the latest developments in aircraft technology displayed during air shows and the Schneider Trophy races. As a result they concentrated on monoplane designs such as the Messerschmitt Bf 109, Heinkel He 111, Junkers Ju 87 Stuka (short for *Sturzkampfflugzeug*, dive-bomber), and Dornier Do 17. All of these performed well in the Condor Legion, which served with the Fascist Nationalists during the Spanish Civil War of July 1936 to March 1939. Göring wanted to test his 'young Luftwaffe' and his aircraft. During the conflict the Luftwaffe gained valuable experience of strategic bombing and discovered that their older Heinkel He 51 biplane was outclassed by newer monoplane designs. It was swiftly switched to a ground-attack role and later saw service as a trainer.

Intelligence of these activities filtered back to Britain, and some of the more thoughtful leaders began to prepare for war.

Times had been hard at the Bentley company. By July 1931 W. O. Bentley had worked through three fortunes, including his own, and Bentley Motors went into voluntary liquidation. Napier had conducted open and proper negotiations to buy the business, but Rolls-Royce, posing as the British Equitable Central Trust to avoid inflating the price, snatched the company away at the last minute. It was, as Setright tells us, 'an unpleasantly furtive Rolls-Royce manoeuvre', and once again it was to lead to bad blood between the aero-engine makers later when Napier was struggling to sell its Sabre engine. Rolls-Royce were powerful in Parliamentary lobbying too: they could summon the aid of 50 MPs, whereas it is doubtful whether Napier could call on even one. Again, Charles Rolls's bequest of prestigious connections was proved beyond price. After acquiring Bentley in this way, Rolls-Royce then started to develop a new Derby-built Bentley 3½-litre car, 'the silent sports car'.

Winston Churchill had an indirect but vital influence on the development of the Rolls-Royce Merlin, but it is hard to see the truth about this remarkable man through the fog of history. He exposed Germany's secret rearmament, he predicted that the situation in Europe would lead to war, and he warned of the weakness of Britain's air power. No other British politician of the modern era can now match his worldwide reputation; he was eventually proved right about German preparations for war, he provided matchless leadership for his nation at her most perilous time of need, and the historian A. J. P. Taylor, not known for his compliments, called him 'the saviour of his country'.[4] So why were his warnings ignored?

Churchill's fate during the 1930s was to echo that of Cassandra, the daughter of King Priam who was cursed by the gods always to foretell the future correctly but never to be believed.

'Throughout the 1930s his was an ancestral voice prophesying war. But his warnings, like Cassandra's, were discredited simply because they were his.'[5]

At the time Churchill's judgement was thought to be poor. During his political career he had crossed the floor in the House of Commons not once but twice, defecting from the Conservative Party in 1904 over the proposed Aliens Bill, joining the Liberals, then later rejoining the Conservatives to become Chancellor of the Exchequer. As a result, he was seen as an opportunist. As he said himself, 'anyone can rat, but it takes a certain ingenuity to re-rat.' During the First World War he was largely responsible for the disaster of Gallipoli, and as Chancellor of the Exchequer he was directly responsible for the return to the gold standard which brought depression, unemployment and the general strike of 1926.

He was the only major politician to support Edward VIII during the abdication crisis, and his views on Indian independence were considered old-fashioned. He described Mahatma Gandhi as a fraud, 'a seditious Middle Temple lawyer, now posing as fakir of a type well known in the East, striding half-naked up the steps of the Vice-regal palace to parley on equal terms with the representative of the King-Emperor'. He suggested that Gandhi should be trampled under the feet of an elephant ridden by the Viceroy.

These views were considered reactionary at a time when most people in Britain thought Indian independence was inevitable. When in 1931 Ramsay MacDonald formed the National Government, Churchill was not invited to join the Cabinet, and he was left in the political wilderness as a backbench MP. And as if to rub it all in, his improvements to Chartwell and his expensive habits of wine and cigars caught up with him, and caused him to fear bankruptcy. As a result he had to spend much of his time writing a biography of his illustrious ancestor the Duke of Marlborough, who as a leader of the allies had foiled Louis XIV's plans to dominate the Continent and whose victories enabled his country to rise from a minor to a major power. It must have galled Churchill to sit writing about this man of destiny, his own ancestor. The contrast may explain where he found the energy and determination to eventually lead his country through the war.

In 1932, Churchill undertook a road trip through Europe with his friend Professor Frederick Lindemann, professor of physics at Oxford, and they were both disturbed by what they saw. 'A terrible process is astir,' Churchill said. 'Germany is arming.' During his wilderness years Churchill sat as a backbencher on the Committee for Imperial Defence and Lindemann sat on an advisory technical Committee for Air Defence Research, which enabled him to feed privileged information to Churchill, who duly aired it in the House of Commons. However, Lindemann was a prickly character, and his fellow committee members grew tired of him. He had a number of odd ideas, one of which, incidentally, was the use of Aerial Mines.

My father had two of these anti-aircraft devices on his Landing Craft (Tank) during the invasions of the Normandy beaches and the island of Walcheren in 1944. They were indeed lethal – to the operators. They involved a rocket, a 122-metre (400-foot) length of wire, a bomb and two parachutes. The idea was to shoot them up in front of attacking aircraft, but of course the unexploded bombs had to fall somewhere, usually on the ship that had just fired it. They were considered so dangerous by the other LCTs that my father made sure the firing buttons were heavily painted over. Another one of Lindemann's aerial-mine ideas was the Holman Projector, a suicidal device which required the crewman to remove the pin from an ordinary Mills bomb grenade, drop it in the steam mortar and hope that (a) the steam valve worked and (b) the grenade didn't fall back down onto him. Thankfully this had a brief life. Some steam trawlers were said to be firing potatoes at German planes for want of Mills bombs. Possibly as a result of the aerial mines and his preference for infrared beams instead of radar, Lindemann was eventually sidelined from the Air Defence committee, but not before his ideas on rearmament began to take hold.

When Churchill first began to raise the alarm about German rearmament, he was not taken seriously by other MPs, partly because they didn't want to hear about it, and partly because he lacked the figures to back up his claims. He soon remedied this by gathering information covertly from concerned officials. He was tireless in reminding Parliament:

Germany is already well on her way to become, and must become, incomparably the most heavily armed nation in the world and most completely ready for war. We cannot have any anxieties comparable to the anxieties caused by German rearmament.[6]

And four months later he was more graphic in his warnings:

… the crash of bombs exploding in London and the cataracts of masonry and fire and smoke will warn us of any inadequacy which has been permitted in our aerial defences. We are vulnerable as we have never been before … This cursed, hellish invention and development of war from the air has revolutionised our position. We are not the same kind of country we used to be when we were an island, only 20 years ago.[7]

However, Churchill himself had no specialised knowledge of aero engines, or indeed any military technology, and some of his ideas were distinctly odd. Although as First Lord of the Admiralty he had encouraged the first use of the British-invented tank in the First World War, by the 1930s he seemed to think that tanks were outmoded. He also thought that fighters would be of little use against bombers and seemed to imagine that aircraft posed little threat to shipping.

Despite these misapprehensions, like him or loathe him, any objective historian surely has to confirm it: Cassandra was right.

Henry Royce was right about the coming war, too, and he was prepared to bet his company's money on it. However, the PV12 did not have an easy gestation, Rolls-Royce made a number of mistakes with the design, and the Merlin, as it soon became known, was not a success at first.

Royce decided to build an engine similar to the Kestrel but bigger, suitable for the next generation of fighters that would need more power. The Buzzard and the R had the power, but they were too big and heavy for the new monoplanes that were coming from Hawker and Supermarine. Incidentally, it is interesting to note that the Griffon, a new 37-litre engine, was first run on 6 January 1933, nine months before the PV12 first ran. This first Griffon was essentially a de-rated R engine with a smaller supercharger, and a much-changed Griffon was produced in 1939. This would eventually replace the Merlin. For now, though, there was no aircraft application in sight for it, and all attention focused on the smaller engine.

So Royce's designers got their pencils and rubbers out and drew an engine that was an almost exact scale-up of the Kestrel, but 20 per cent bigger in capacity: 27 litres. Then to give the pilot of the aircraft a better view they turned it upside down! The reason for this is that a V-configured engine is narrower at the bottom than the top. Also the exhaust ports can exit below the fuselage, improving the cooling. A wooden mock-up of this upside-down engine was made, and this was the cause of one of the most bizarre stories of inadvertent industrial espionage of the Thirties. A group of German aeronautical engineers was touring the Rolls-Royce factory and glimpsed the upside-down mock-up on Arthur Rowledge's office floor:

The Germans evidently thought they had noticed something of supreme significance. There is every reason to believe that the design of the inverted Daimler-Benz engine used in the Messerschmitt 109 and the Junkers engine sprang from this visit to Derby. From their point of view, the inverted engine was desirable because it enabled them to fire the cannon through the airscrew shaft, but this had the serious result of forcing them to mount the supercharger on the side of the engine instead of at the end, a position which necessitated complex piping and which made it difficult to find a suitable place for carburettors. The Germans' later preference for direct fuel injection was attributable to the difficulty of carburettor layout, and not to any objection to carburettors as such.[8]

The visiting engineers had clearly taken notice of Ovid's dictum: *Fas est ab hoste doceri*: one should learn even from one's enemies.[9] It seems hard to believe now that potential enemies were allowed to roam freely around the most secret new designs (couldn't other ridiculous designs have been left lying around the workbenches like the pigeon-guided missile, or the anti-tank dogs? Except that those were real enough).*

This is another one of the Rolls-Royce stories that appeared in the company's official history, so presumably the directors believed it. However, a Daimler-Benz historian might point out that their V12 engine had already been turned upside down well before this factory visit: in fact, as we have seen the 1913 Mercedes aero engine featured this configuration. The pilot's-view theory doesn't hold water with the German twin-engined wing-mounted instal-lations, which were also inverted.

To explain the direct fuel injection mentioned above: it's rather like fuel injection in a diesel engine. Direct fuel injection directs petrol at high pressure through a nozzle (injector) mounted inside the combustion chamber, instead of the fuel being mixed in a

---

* 'Project Pigeon' used pigeons as pilots in US glider-bombs. They were trained to peck at a screen showing an image of the target, steering the missile. The Russian anti-tank dogs carried mines and were taught to run under enemy tanks. A stick-fuse was triggered by contact with the hull. Trained on diesel Soviet tanks, the dogs followed their noses and ran towards their own tanks instead of the petrol-powered German tanks. Those that survived, terrified by the noise of the engines, ran back to their handlers with predictable results.

carburettor with the air before the supercharger, then introduced through the inlet valves. Rolls-Royce preferred the latter method because the evaporation of the petrol resulted in the fuel/air charge being considerably cooler. The main advantage of direct fuel injection was that in a negative-G dive the German engines would still run at full power, whereas the carburettor-fed Merlins would momentarily cut out as the fuel surged. A quick-fix solution to this was dreamed up by a young woman, Beatrice 'Tilly' Shilling, as we shall see later.

When presented with the upside-down PV12 wooden mock-up, the established airframe manufacturers Supermarine and Hawker told Rolls-Royce they had better rectify their first mistake and turn it upright again. Certainly it would have changed their fuselage shapes to a degree, but the fact remains that visibility was seriously hampered on the Merlin-engined Hurricanes and Spitfires, particularly when taxiing. A snake-like mode of progression had to be adopted; the pilot peering anxiously either side of the enormously long and wide engine.

Rolls-Royce made a number of odd mistakes during the design of the Merlin. They chose to depart from the design of the excellent and well-tried Kestrel, whose only faults were that it was five years old in a fast-moving field and not large enough for the work to come. One reason could have been Royce's health, which was deteriorating fast. He had worked until the end, the 70-year-old sketching a design on the back of an envelope. It was for an adjustable shock absorber for the company's new Bentley 3½-litre sports car. He gave it to his nurse and housekeeper, Ethel Aubin, to pass on to the 'boys at the factory'. His sketch didn't reach the factory until he was dead, because he died the following day, 22 April 1933, at his home, Elmstead House, West Wittering. It is curious to note that he had never travelled in an aeroplane.

'The man departs. There remains his shadow.'

Henry Royce left most of his £112,000 (£8m) estate and his home at West Wittering to Ethel Aubin, but there was also a smaller bequest for his wife. It is telling that he left his ashes in his nurse's care: she had been more than a wife to him. Ethel married Royce's solicitor George Tildesley in 1935, but they separated in 1937 when his homosexuality became apparent. Ethel had been a vital part of the Rolls-Royce story, and she survived until 1965.

The design office was uprooted from West Wittering, and now the ex-Napier engineer Arthur Rowledge was in poor health. The loss of both Royce's and Rowledge's experience may have led to mistakes in the design of the new PV12 aero engine, which had to be corrected by remorseless testing. Nearly all the departures from the Kestrel design had to be reversed in the end. The commonplace myth that the Merlin was a perfect design is far from true: it was the result of repeated failures and relentless problem-solving.

When the PV12 engine was first run in October 1933, the first problem was of the double-helical reduction gears breaking up. As we have seen, one of the Rolls-Royce engine's advantages was in the use of a speed-reduction gear between the engine crankshaft and the propeller. This ensured that the engine could run fast enough to make maximum power, and the propeller could run slow enough to be most efficient. The first gears used on the PV12 were double-helical, which are in a herringbone pattern with a groove between the rows of teeth. If you have ever glanced at a Citroën logo and wondered what the strange double chevrons represent, they are a tribute to the gears that made the company's fortune.

When visiting his mother's birthplace in Łódź, Poland, the 22-year-old André Citroën had seen a carpenter cutting a fishbone shape into a large pair of wooden gear wheels intended for a mill.

They ran quietly and efficiently. He bought the patent for very little money and put them into production for marine steam turbines. They overcame the sideways loads of ordinary helical gears (cut at an angle), which would exert enormous axial pressure. Citroën became a major armaments manufacturer and in 1934 produced his visionary *Traction Avant* car, which with its front-wheel drive, all-round independent suspension and monocoque body showed the way for the next 100 years. Development costs of such a leap forward were crippling, and Citroën declared bankruptcy in late 1934. The double-chevron logo continues, but few owners know what it represents.

Meanwhile back in Derby, testing of the PV12 continued painfully. The double-helical reduction gears had been replaced by stronger straight-cut spur gears, but now the cylinder jackets were cracking. This was because Rolls-Royce had adopted an all-in-one monobloc construction for the aluminium cylinder blocks and the upper half of the crankcase (this integral construction was used for the Phantom III car engine). Arthur Rubbra, one of Rolls-Royce's engineers, tells us:

In order to provide a more rigid engine crankcase, to allow for higher crankshaft speeds, it was decided to cast the cylinder jacket portion of the cylinder block in one piece with the crankcase and to provide a separate cylinder head with the cylinder liner clamped between it and the crankcase. This presented quite a foundry problem in maintaining sectional thickness throughout, but this was solved in due course. It was soon discovered, however, on development that a major difficulty was presented because failures in the reciprocating components usually resulted in serious damage to the large crankcase-cum-cylinder jacket casting, this proving an expensive and time-consuming job.[10]

This is not the whole story. Rubbra is suggesting here that the reason for changing the design was that if the PV12 threw a connecting rod out the side it would write off not only the upper crankcase casting but also the two cylinder blocks, as they were all in one piece. In fact there was a far more serious problem here which was being airbrushed out of history. The more objective Major George Bulman, who was an excellent member of the Air Staff, instead reported:

> Initial Merlin production started well after several type tests were run, but after about 100 engines had been made an epidemic of cracks in the walls of the aluminium combined crankcase and cylinder blocks developed. Hives and ourselves had a desperate investigation into the casting procedure but after an agony of indecision for a few days we decided to literally cut the Gordian knot by splitting the one-piece casting into three: crankcase and two cylinder blocks! Frantic tests of the new construction were hurried through, and the trouble disappeared. Production with the drastically modified construction restarted, and thanks to the inevitable setbacks in the output of the first Battles, Hurricanes and Spitfires, Rolls were able to regain their substantial lead in Merlin deliveries to meet aircraft output. But it was a harassing few months, peace mercifully prevailing!

Peace was indeed still prevailing, largely thanks to the delaying tactics employed by Britain's Prime Minister, Neville Chamberlain. It is more than likely that he knew just how unprepared Britain was for war, and he thus deliberately adopted a conciliatory approach to Adolf Hitler.

Meanwhile, things were getting worse in the PV12 test cells. The prototype engine eventually passed a type test in July 1934, giving just 625 hp at sea level and 790 hp at simulated 12,000 feet (3,657 metres), but this was barely more than the Kestrel, and in some circumstances less.

The PV12 first flew in a Hawker Hart biplane (serial number K3036) on 21 February 1935. It used evaporative cooling, a system then briefly in vogue. This allowed the water coolant to boil, the idea being that the phase change from water to steam removed more heat from the engine, thus allowing less coolant to be carried. The problem was that the wing-mounted condensers needed to return the steam to water were bigger than radiators, and were therefore more vulnerable to enemy fire. What seemed to be a good idea bit the dust.

When ethylene glycol coolant became available from the US the engine was switched to this sweet-tasting (and highly toxic) stuff now found in anti-freeze. With a lower freezing point than water and a higher boiling point, it was ideal for aircraft engines, but as the compound was so flammable (it was used in dynamite) it was another source of fire. With ethylene glycol it was possible for the engine to be run at much higher temperatures, and now a much better idea came along: a more conventional high-pressure, high-temperature system using coolant under a pressure of 30 lb per square inch. This allowed the temperature of the coolant to increase from 80° Celsius (176° F) to eventually as much as 140° Celsius (284° F), and that meant that, because of the steeper temperature gradient between the cooling air and the radiating surfaces, the radiators could be half the size. The smaller the radiator, the less the drag, and the less drag, the faster the aeroplane.

In an attempt to increase the power output a new cylinder-head design was tried, the so-called ramp head. This had been designed by senior designer Albert Elliott for one of the Rolls-Royce cars, using just two valves and a combustion-chamber shape which imparted a lot of turbulence to the fuel/air charge and thus improved combustion. He had to adapt it to the four valves of the aero engine. The inlet valves were set at 45° to the top of the cylinder block and the two flat planes in the combustion-chamber roof were of unequal width and angled, so the design was dubbed the 'ramp head'. It proved to be another failure.

The engine was now being known as the Merlin B, and it gave 950 hp. The new ramp head was kept on all experimental Merlins until the F, which went into production as the Merlin 1. However, the new cylinder head was giving all sorts of new problems: exhaust valves were now distorting and failing, detonation was occurring, and castings were cracking. Arthur Rubbra was in the test cell:

> The improvement in detonation characteristics was not felt on this engine, being worse than with the normal head, and a further trouble presented itself on the type test when a hole was burned through the exhaust port wall into the coolant jacket space causing a spectacular fire of the glycol coolant.[11]

Not content with bursting into flames, now the engine's valve gear was breaking up. New asymmetrical cam lobes helped with this problem. At least the crucial crankshaft and connecting rods were reliable. The Merlin's strong bottom end was what enabled the power to be more than doubled through the engine's life.

Now the Hawker Hurricane fighter and Fairey Battle light bomber production lines were tooled up and ready for the new engine. But it was nowhere near ready.

The Merlin C was given separate crankcase, cylinder blocks and bolt-on cylinder heads to solve the cracking problems after the B had run just 50 hours. When it was ready, flight testing began in a

Hawker biplane in 1935, with 950 horsepower at 11,000 feet (3,400 metres).

Yet another problem now raised its ugly head; glycol is nasty, insidious, searching stuff, and it found any leaks that it could – and escaped. As we have seen, this was a continual problem with the Kestrel, it was duly passed down to the Merlin, and the engineers found the overheating engines were distorting their cylinder heads. The engineers came up with another 'fix': they developed a cooling system that was kept at a high pressure of around 2.8 bar (40 lb/in). Their 70 per cent water and 30 per cent ethylene glycol mixture was still able to reach 130° C without boiling.

The Merlin C failed its 50-hour type test in May 1935; a later Merlin passed it in December but failed the 100-hour military test the following March. The pressure was coming onto the engineers to get this engine right. What they did was to go right back to the Kestrel design and redesign the cylinder head and cylinder block in one piece. The valves would now be parallel, but no doubt fingers were crossed in the test cells in Derby. The result was called the Merlin G, but Rolls-Royce's problems were not over. They now had doubts about this new integral Kestrel-type head because there was precious little space to provide for a seal between the cylinder liner and the cylinder head. Sure enough, coolant leaks started appearing at the join. So the development engineers designed yet another new cylinder block with a separate cylinder head. At last this solution seemed to work, but Rolls-Royce could not change to this design until 1942, when the Merlin 61 was introduced. The Packard company in America were actually the first to make this improved design when they were licensed to produce the Merlin. So another problem had been solved – eventually. This was a difficult and painful birth.

The Rolls-Royce philosophy of remorselessly testing eventually got results, but it also suggests that incorrect design work had been done at the drawing-office stage. Henry Royce must have been spinning in his grave at 3,000 rpm.

The pace of technological evolution is staggering. Humans started making stone tools around 2.6 million years ago: stone hammers and sharp stone flakes; 200,000 years ago they had only progressed to pointed tools and scrapers. The wheel (or axle) came along 3,500 years ago, but it took well over 3,000 years more to come up with the steam locomotive. But by the 1930s the pace had increased exponentially, and Rolls-Royce had taken the aero engine from 12 horsepower to over a thousand in half a lifetime. Soon, in four years of wartime, they would double that figure.

How was this possible? The Renaissance concept of *Homo faber* (man the maker) explains how human beings can control their environment by the use of tools and thus control their fate: *homo faber suae quisque fortunae* – every man is the creator of his destiny. But humans such as Henry Royce could also pass on knowledge between generations with language, drawings and training. Today, with contemporary advances in artificial intelligence and the skyrocketing of progress, a pessimistic historian could see the end of humanity in a matter of a thousand years.

Now war was in sight, and Rolls-Royce was being pressed hard to provide engines for Fairey Battle bombers and Hawker Hurricane fighters. So they persuaded the Air Ministry to relax the type-test requirements so that they could replace the failing valves during the test, and went into production in the Derby factory with the Merlin F so that aircraft could actually be built. Even so the Merlin Mark 1 production engine, as it was called, could not get through this relaxed test until November 1936, five months after delivery of the first engine. All Rolls-Royce could do was to tell Hawker to wait for the production version of the G, the Merlin 2, and deliver the 180 Mark 1 engines they had already built to the Fairey company for their Battle bomber. In August 1937 the Merlin 2 was in production, weighing 1,375 lb and developing 1,310 hp. Rolls-Royce could now get on with what they really excelled at: making their engine better.

Between the wars the far right in Europe had grown bolder, with the rise to power of the journalist and politician Benito Mussolini in Italy in 1922. Then came the 1933 election of Adolf Hitler to the position of Chancellor of Germany, and then the accession of Francisco Franco, who became dictator of Spain in 1939 after his victory over the Republican government during the Spanish Civil War. Popular nationalism was on the rise across Europe, and so war was clearly on the way.

CHAPTER THIRTEEN

Guernica

Guernica is a quiet town in the Basque region of northern Spain. I went there to find out about one of the most significant events of the Spanish Civil War. A few hundred metres along from the church is a tiled copy of Pablo Picasso's *Guernica*, the anti-war work he painted after hearing what happened here. If you sit on the nearby bench you can study it and try to decipher the meaning. It is a jumble of wounded people and animals, a claustrophobic mass of violence and suffering. There is a gored horse, a bull, screaming women and dead soldiers.

The 26th of April 1937. 'It was market day, and there were finally some sweets on sale once again,' said Luis Iriondo Aurtenetxea, who was a 14-year-old boy on that day. Then aircraft were seen flying up the estuary and the horror began.

This was a dress rehearsal for the Blitzkrieg, or 'lightning war'. The aircraft were from the Nazi German Luftwaffe's Condor Legion and the Fascist Italian Aviazione Legionaria, operating at the behest of General Francisco Franco's nationalist government. The Nationalists regarded Guernica as a communications centre for their Republican enemies. The German squadron leader Wolfram

German advertisement for the Junkers Ju 87 Stuka dive-bomber.

Freiherr von Richthofen, a cousin of the 'Red Baron', had planned the operation meticulously. A Monday was chosen to maximise the number of civilians packing into the town on market day. After unloading around 20 tons of incendiary, splinter and blast bombs, Richthofen noted that the town 'has been literally razed to the ground. Bomb craters can be seen in the streets. Simply wonderful!'

Things were not so wonderful on the ground. Although the victims could see the pilots' faces, there was no attempt to avoid dropping bombs on the crowds of women and children. Between 170 and 300 people died, and hundreds were wounded by bomb splinters or by fire. Fighter planes strafed people running away across the fields. The boy Iriondo survived by darting into a cramped shelter beneath a terrace filled with other townspeople. 'We thought we would suffocate. One of us tried to light a match, but there wasn't enough oxygen.' Inside he cowered under the onslaught for around half an hour until there was a pause in the bombing. Then the townspeople staggered out into the carnage and thick smoke, unaware that the bombers had gone to Vitoria to load up with fresh bombs. Many residents who survived the first attack died in the second, unaware of the returning aircraft.

By the evening the bombers seemed to have gone and the boy finally crept out of the shelter. His town, which had once been the cultural centre of the Basques, was in flames. Guernica had practically ceased to exist. And, typically of area bombing, the supposed targets of a weapons factory and a bridge had escaped completely unscathed.

At the news Picasso immediately abandoned a mural project for the Spanish Republican government and started on a new painting. He used black, white and grey paint and sharp edges which bring to mind a collage of newspaper atrocity photographs. We seem to be in a chamber like the one that sheltered Iriondo, giving a feeling of claustrophobia and chaos. A bull stands over a grieving woman and a horse falls wounded in agony. Dismembered bodies lie around and a woman floats through the window, witnessing the scene by the light of an oil lamp. A soldier lies on the floor, his open palm disclosing stigmata.

As is often the case with Picasso, the work does not lend itself to easy interpretation. When quizzed on why the painting included a bull and a horse, he himself said: 'This bull is a bull and this horse is a horse … If you give a meaning to certain things in my paintings it may be very true, but it is not my idea to give this meaning. What ideas and conclusions you have got I obtained too, but instinctively, unconsciously. I make the painting for the painting. I paint the objects for what they are.'

It is worth pointing out that the bull and the horse often represented Spain in Picasso's work. In sketches done at the same time as *Guernica*, *The Dream and Lie of Franco*, he depicts Franco as a monster that first eats his own horse, then kills a bull. *Guernica* received little attention when first exhibited in Paris, but now it is probably Picasso's most famous work, and certainly his most direct anti-war message. The original was literally nailed up in a Manchester car showroom for two weeks to raise support for the Republicans: it was seen as an anti-war banner.

While Picasso was living in Nazi-occupied Paris during the Second World War, a German officer studied a photo of *Guernica* in his apartment. 'Did you do that?' he asked. 'No,' Picasso replied, 'you did.'

The bombing of Guernica crystallised the fears of many in Britain. An influential book of the Twenties, *The Command of the Air*, argued that air power had changed the rules of warfare: no longer would armies face each other across battlefields; bombers would usher in an era of total war where civilian populations would be the victims of strategic bombardment. 'The bomber will always get through' was a disturbing phrase used by the Conservative Party leader Stanley Baldwin in a speech to the House of Commons:

> I think it is well also for the man in the street to realise that there is no power on earth that can protect him from being bombed. Whatever people may tell him, the bomber will always get through. The only defence is in offence, which means that you have to kill more women and children more quickly than the enemy if you want to save yourselves.

In this speech Baldwin argued against unilateral disarmament and he clearly showed a preference for bomber aircraft as an offensive deterrent instead of fighter aircraft employed in defence.

Either way, these aircraft would need powerful and reliable aero engines, and on 12 April 1935 the improved PV12 engine no. 1 flew again in the Hawker Hart biplane at the Aeroplane and Armament Experimental Establishment at Martlesham Heath. Most subsequent testing took place at Rolls-Royce's own flight-test establishment at Hucknall aerodrome, 20 miles from the Derby factory. These biplanes were not ideal for testing the new engine. Rolls-Royce test pilot Ronald Harker takes up the story:

> We found that the drag of these biplanes was such, that hoped for improvements in drag reduction could not be measured and so it was decided to obtain one of the cleanest aircraft available so that we could note the results of such improvements as smaller radiators, rear facing exhausts etc. The Heinkel 70 was chosen, and a Kestrel engine was sent over to Rostock in Germany for installation and to replace the BMW engine. The

Heinkel was a four-seater passenger and mail plane and was by far the most advanced and elegant aircraft available. The story goes that before they installed the Kestrel into the Heinkel, they put it into their own Me 109 and possibly a Heinkel 113 fighter for their own edification![1]

With the Kestrel installed, the Heinkel would do 300 mph, which was 60 mph faster than the best British fighters. Soon the Rolls-Royce engineers could measure the benefits of more aerodynamic radiators, and rear-facing exhaust stubs, which would add 11 mph to the future Spitfire.

Intriguingly, Harker's quote illustrates Germany's duplicity: as we have seen the Heinkel He 70 was secretly intended to be used as a light bomber but appeared to be a passenger and mail aeroplane, and the Heinkel 113 fighter mentioned by Harker was an entirely fictitious aircraft, invented by the Nazi Minister of Propaganda Joseph Goebbels to impress either foreign powers or the German populace. The second German plane to be tested with the Kestrel was the Junkers Ju 87 'Stuka' dive-bomber, which was intended to be powered by the engine. Ten Rolls-Royce Kestrel engines had been ordered by Junkers on 19 April 1934 for £20,514, 2s and 6d.[2] Presumably they wanted the later single Kestrel as a comparison to ensure they had not been short-changed with lower-powered versions.

However, the prototype Stuka crashed on 24 January 1936 at Kleutsch near Dresden, killing Junkers's chief test pilot, Willy Neuenhofen, and his engineer, Heinrich Kreft. This did not impress the Reichsluftfahrtministerium (RLM), the German aviation ministry, and nor did the fact that it used a British engine. Jumo engines eventually replaced the Kestrels, and the Stukas with their terrifying Jericho-Trompete (Jericho trumpet) wailing sirens devastated Spanish Republican targets. Later this sound dominated the propaganda films of the blitzkrieg era and demonstrated how the Nazis used terror to crush their victims on the ground.

There are two strange stories about the Stuka. In May 1938 three aircraft were used in Spain to evaluate their killing abilities, being sent to bomb four villages in Castellón well away from the conflict: Albocàsser, Benassal, Vilar de Canes and Ares del Maestre. Thirty-eight villagers were killed and the village centres destroyed. A few days afterwards several Germans came to the villages to photograph and evaluate the damage. It appears that the purpose of this exercise was to test the Stuka and the aircrews.[3]

And in 1939 in preparation for the invasion of Poland a mass-bombing demonstration by Stukas was put on for a group of high-ranking Luftwaffe officers. The idea was to burst through low cloud and simultaneously release their practice bombs, causing much amazement. Thirteen Ju 87 Stukas and their crews dived through the cloud at great speed, sirens wailing, expecting any moment to emerge below the cloud ceiling, release their bombs and pull out of the dive. However, no one told them that a light mist had crept in and the low cloud now extended to the ground. They slammed into it, killing all 26 crew and destroying all 13 Stukas.[4]

The death of test pilots was an accepted fact of aeroplane development. In the 1950s it was said that test pilots were being killed at the rate of one a week. They were a special breed of highly intelligent yet courageous individuals (qualities one might think were mutually exclusive). The test pilot had to understand a test plan and then fly the aeroplane in a specific way, sticking rigorously to the plan. They had to have a sensitive feel for the aeroplane, and sense why it might be behaving oddly. They might have to cope with a multitude of things going wrong and try to stop them cascading into disaster. On landing, they had to explain their observations to engineers who might not know how to fly a plane.

The Messerschmitt test pilot Fritz Wendel knew there was something terribly wrong with the Messerschmitt 210 V1. One thousand examples had been ordered off the drawing board, but the first prototype had crashed. Wendel was ordered to test the second prototype and he took off with misgivings. The Me 210, he thought:

was nearly as bad as Heinkel's He 177, surely the worst plane we produced in Germany throughout the war years. I was never happy about the Me 210. I have only been forced to bale out of an aircraft twice, and the first time was from this twin-engined fighter-bomber.

It was on 5 September 1940, that I took off from Augsburg-Haunstetten in the Me 210V-2. We had more than a suspicion that the tail assembly was weak, and that there was a strong possibility that it would part company with the rest of the airframe in a dive. I was flying solo, the mechanic who usually occupied the radio operator's seat on test flights being left on the ground as this flight was likely to be more than usually dangerous.

During the climb I recall the thought flashing through my mind: 'If something goes wrong, remember to avoid baling out over the forest.' At 9,000 feet I checked my instruments and started the planned series of dives, my unpleasant thoughts temporarily forgotten. Stick forward ... down went the nose and the needle began to creep round the airspeed indicator. Stick back, and the nose lifted. Stick forward ... back ... forward, and so on. One more shallow dive and back to the airfield. A thousand feet showed on the altimeter and I began to level off. Then it happened, and right over that blasted Siebentischwald again! The plane shuddered, the tail fluttered, and bang, the starboard elevator broke off! Immediately the plane went into a half-loop downwards. Before I could gather my wits I was flying on my back in the direction from which I had just come. I knew exactly what would happen now, as I had seen the same thing happen to the first prototype some weeks before. The plane would fly inverted for several seconds and then dive straight into the ground.

Hanging on my seat straps, head downwards, I automatically grabbed the release pin, but I hadn't jettisoned the canopy. In that split second I fumbled for the roof's emergency release lever, pulled it, and the canopy flew off with a bang. I pulled the release pin … the chute opened with a glorious crack. I looked down, and there were those damned trees almost touching my feet. But I was in luck, for the wind blew me towards the only clearing that I could see, and I escaped with nothing worse than a sprained ankle. I had lost a prototype, but at least I had proved its unsatisfactory characteristics and had lived to tell the tale. But there are few things in the world harder than proving to an aircraft designer that his latest pet creation is simply not good enough.[5]

Alex Henshaw was famous for being the first test pilot to barrel-roll a Lancaster bomber. He was the chief Spitfire and Lancaster bomber test pilot based at the Supermarine factory at Castle Bromwich, which produced up to 320 Spitfires a month. He tested up to 20 aircraft a day. He had no fewer than 11 engine failures which pointed to a fatal weakness in the Merlin design and revealed possible wartime industrial sabotage. The Merlin engine would suddenly cut out dead during flight and Henshaw would desperately try to spot somewhere to land. On one occasion this happened over Willenhall, a heavily built-up area of Birmingham, and he landed between two rows of houses in a cabbage patch with the wings torn off. He managed to survive in the cockpit. Understandably he began to ask why there were so many failures.

As with most aero engines, the Merlin had two magnetos to fire the twin sparking plugs in each cylinder. The idea is redundancy; if one magneto failed the other would still provide a spark. The fatal flaw was that both magnetos were driven by one cross-shaft in the wheelcase, a veritable forest of gears. For some reason the bronze angled skew gear that drove this was suddenly being stripped of all its teeth, resulting in what Henshaw used to call 'The Deafening Silence'. Urgent investigations revealed that the Crewe factory had been the source of all the failed engines, and when 81 engines were dismantled 19 were found to have a loose nut on the upper vertical drive shaft. The fitter responsible was prosecuted and received a prison sentence.

Guernica burns after the Blitzkrieg attack of 1937.

CHAPTER FOURTEEN

The Spitfire is typically British. Temperate,
a perfect compromise of all the qualities
required of a fighter, ideally suited to its
task of defence.[1]

The Luftwaffe may have realised that the biplane was obsolete by 1930, but some at the British Air Ministry had not. Older officers remembered the disintegration of two monoplanes back in 1912. They also disliked the higher landing speeds of an aeroplane with only one pair of wings, and they may have recalled their glory days in the highly manoeuvrable biplanes of the First World War. Like most military leaders they were busy fighting the last war and wishing it was the one before that. Either way they were dead against the monoplane type, and no government money was forthcoming.

So Rolls-Royce were not the only company obliged to invest their own money in the defence of their country. Just two years after his attempt to fly across the Atlantic Harry Hawker had been killed testing an aircraft, but his company had survived. Hawker Aircraft Limited's designer Sydney Camm drew a new monoplane

K5054, the Spitfire prototype, designed by Reginald Mitchell and first flown in 1936. Here the elliptical wings are shown to advantage.

fighter that was eventually to be called the Hurricane. Ever since the Battle of Britain this aircraft has been overshadowed in the public mind by the more glamorous Spitfire, but according to Fighter Command the Hurricane accounted for 55 per cent of the 2,739 German losses during the conflict, compared with 42 per cent by Spitfires. The Hurricane was also the first fighter to be designed around the new Rolls-Royce Merlin engine.

Camm, like Reg Mitchell at Supermarine, had realised by the end of the 1920s that the biplane was finished. In the air exercises of 1930 no biplane fighter could catch his light bomber design, the Rolls-Royce Kestrel-powered Hart, whose pilot was instructed to slow down to allow the fighters to catch up. Camm started to work on his ideas for a monoplane fighter at about the same time as Mitchell was starting to sketch the Spitfire.

In 1930 the Air Ministry circulated a draft version of Specification F7/30 for a heavily armed day and night biplane fighter, and Camm submitted a design for a scaled-up version of his successful Fury biplane, dubbed Hawker PV3, another private venture. The Ministry had pushed hard for an experimental Rolls-Royce engine to be used: the steam-cooled Goshawk, which was a development of the Kestrel using an evaporative cooling system. This was a brief fashion and only one Hawker PV3 biplane was built. There was no order for it from the Air Ministry.

This was altogether a good thing, because Sydney Camm then built what he really wanted; a fast monoplane fighter based around the new Rolls-Royce Merlin engine. And, just as Rolls-Royce had to fund their PV12 Merlin, Hawker had to find the money for their new monoplane fighter. Rolls-Royce contributed to the cost of the prototype, K5083, in the hope that Hawker would be a large customer for the Merlin. In this their hopes were met.

Camm squeezed the new engine into his existing fuselage design, moving the radiator back for better weight distribution, and this enabled him to fit a retracting undercarriage. This of course reduced drag and increased speed. He also fitted eight machine guns. Discussions with Ralph Sorley, a young squadron leader (later Air Marshal), had convinced the Air Ministry that the high speeds of monoplane fighters would enable pilots to hold their targets in their sights for no more than two seconds, and this persuaded them to revise their specification to F36/34 to include eight machine guns in the wings. The new aero engine was beginning to search out the limits of the Stone Age human brain.

When Sydney Camm's new 'Interceptor Monoplane' aircraft landed back at Brooklands after its maiden flight on 6 November 1935, Camm climbed onto the wing and quizzed the test pilot. George Bulman pushed back his goggles and congratulated him: 'Another winner, I think.'

In February 1936 the prototype Hurricane was delivered for service evaluation by test pilots and the reports came back in: the Hurricane had remarkable ease of handling at all speeds down to stall. The top speed turned out to be 315 mph at 16,000 feet, considerably higher than the Air Ministry's requirement of 275 mph. However, the new Rolls-Royce Merlin engine was giving trouble.

The test pilot, Philip Lucas, reported:

The only thing which marred the otherwise very satisfactory trials was continued unreliability of the engine. There were at least three engine changes during the first two weeks due to a variety of defects, the most serious being internal glycol leaks causing rapid loss of coolant, coupled with distortion and ultimate cracking of the cylinder heads because of the much higher operating temperatures possible with this type of coolant. Soon it was apparent that the engine required a great deal more development before it became sufficiently reliable for Service operation.

The engineers back in Derby were frantically trying to solve these problems, and in time they did. Just about everything that could go wrong with the protype Merlins had gone wrong, and that could explain why in the end it turned out to be such a thoroughly developed engine.

The Hurricane proved to be a great success once the Merlin was reliable. In June 1936 it was ordered into production by the Air Ministry, and it started to join squadrons on Christmas Day 1937. It was conventionally built in the same way as the previous biplane generation, and it might surprise modern readers to learn that most of the external surface of the Hurricane was made of linen. A skeleton frame was constructed from steel girders and alloy cross-bracing bolted together rather than welded. Screwed onto these were wooden formers and stringers to give a rounded shape, and over these was stretched the linen covering. This was doped or lacquered to stretch and waterproof the material and make it airtight. Aircraft dope was highly flammable nitrocellulose, also known as the explosive propellant guncotton. The children's author Roald Dahl was a flying ace who flew Hurricanes and whose plane exploded into flames when he crashed the similarly built Gloster Gladiator:

They have taut canvas wings, covered with magnificently inflammable dope, and underneath there are hundreds of small thin sticks, the kind you put under the logs for kindling, only these are drier and thinner. If a clever man said, 'I am going to build a big thing that will burn better and quicker than anything else in the world,' and if he applied himself diligently to his task, he would probably finish up by building something very like a Gladiator.[2]

To speed up production the first Hurricanes had fabric-covered wings based on earlier Hawker designs, but metal wings were substituted in late 1939. When the Hurricane was introduced its construction seemed old-fashioned compared with other designs which had the new stressed-skin all-metal construction. However, it was easier and cheaper to build, requiring only 10,300 man- or woman-hours instead of the 15,200 it took to build a Spitfire. It could also be repaired in the field in ways the more sophisticated Spitfire could not.

Hawker Hurricanes were also manufactured in Canada by women on production lines overseen by a woman. Hawker, realising that war was inevitable after the Munich Crisis of 1938, arranged for the building of Hurricane airframes at the Canadian Car and Foundry in Fort William, Ontario. Rolls-Royce Merlin engines would be fitted once the airframes were shipped to Britain. The factory's chief engineer was Elsie MacGill, who became known as 'Queen of the Hurricanes'. She was the first Canadian woman to earn a degree in electrical engineering, but just before graduating she had contracted polio and was told that she would spend the rest of her life confined to a wheelchair. Instead she learned to walk supported by two metal sticks. When her factory was selected to build the Hawker Hurricane fighter aircraft her role was to oversee and smooth the production, and half of her 4,500 workers were women. MacGill was celebrated in a comic book biography after the Canadian Car and Foundry had produced a total of 1,400 Hurricanes. Neither her gender nor her disability had stopped her from using her considerable talents to help win the war.

During the early 1930s Britain had expected any future war to be fought with France, not Germany.[3] This is why the Air Ministry's Specification P27/32 demanded a light bomber that could carry

1,000 lb (450 kg) of bombs at a speed of 200 mph (320 km/h) over a distance of 1,000 miles (1,600 km). This range would be sufficient to reach Paris from airfields in England, but too short to reach Berlin – and return. Another miscalculation was to specify a light bomber in case the 1932 Geneva Disarmament Conference banned heavy bombers. In the event, Germany withdrew from the Geneva Conference and the League of Nations when Hitler came to power in 1933. As a result the Fairey Battle bomber was obsolete before it was built.

The Battle was the first production aircraft to be equipped with the Merlin engine, and at first it all looked good. The new Merlin was rated at 1,030 hp, a greater output at that time than any other engine in the world. The Merlin had a small frontal area so that the Battle was able to be designed as an elegant sharp-nosed all-metal single-engine aircraft, with a low-mounted cantilever monoplane wing and even a retractable tailwheel undercarriage. As a result it was aerodynamically slippery and could fly at 257 mph, far faster than the 185 mph achieved by the Hawker Hart and Hind biplanes it replaced. It wasn't going to be fast enough. The Battle had a long transparent canopy for the three crew members and so it looked rather like a lengthened single-engined fighter aeroplane. However the crew, the bombs and a great deal of heavy equipment were all just too much for the single Merlin.

In contrast, consider the Supermarine Spitfire, perhaps the most famous piston-engined aircraft ever, and certainly one of the most beautiful weapons of war. Much has been written about the Spitfire, but perhaps not so much about its relationship with the Rolls-Royce Merlin engine.

We have already met Reg Mitchell, the Spitfire's designer. After the huge success of his Schneider Trophy racers he had gained a reputation as one of the world's most innovative aircraft designers. After all, his Supermarine S6B monoplane had broken through the 400 mph barrier to set a new world speed record of 407.5 mph, and this was in an era when the best RAF biplane fighters such as the Gloster Grebe and Armstrong Whitworth Siskin struggled to exceed 150 mph.

So when in 1931 the Air Ministry issued Specification F7/30, Mitchell's design team thought they could easily meet the conditions. It called for a fighter with a minimum speed of 195 mph in level flight at 15,000 feet (4,572 metres), to be made entirely of metal, mounting four machine guns, enabling good visibility for the pilot, having a landing speed of 60 mph and a service ceiling of at least 28,000 feet (8,534 metres). At last some in the Air Ministry were trying to breathe fresh air into the aircraft industry.

Beverley Shenstone was a Canadian aerodynamicist who worked with Mitchell and his team throughout the Spitfire development. He looked back at Supermarine's first attempt to meet F7/30:

> My own feeling is that the design team had done so well with the S5 and S6 series of racing floatplanes which in the end reached speeds of over 400 mph that they thought it would be child's play to design a fighter to fly at little over half that speed. They never made that mistake again.[4]

The result was the first failure of Mitchell's career: the Type 224. It was an ugly brute with huge inverted-gull wings, an open cockpit and a large speed-sapping fixed undercarriage with baggy 'trouser' fairings. These contained plumbing for the Rolls-Royce Goshawk engine, which, as we have seen, was an engine that used the briefly fashionable evaporative cooling system. The trousers contained tanks collecting condensed water which was then pumped back to the engine. The whole contraption was decidedly lacking in the streamlining that had led to the success of Mitchell's racers.

One possible reason for Mitchell taking his eye off the ball was his health. He had been feeling unwell for months and was diagnosed with rectal cancer in 1933. He was operated upon to route his colon outside his abdominal wall. As we have seen, like Henry Royce he would have to wear a colostomy bag for the rest of his life. And like Royce, he was tough and managed to do some of his best work with this disability. Unusually for an aircraft designer in those days, he devoted the time and effort to take flying lessons and gained his pilot's licence in July 1934.

Sir Hugh Dowding, the Air Member for Supply and Research, believed that a sophisticated defence network, fronted by the best fighters, *could* prevent bombers getting through. He was distinctly unimpressed with the Supermarine Type 224, having hoped that Mitchell would have come up with something akin to the S6. He was particularly irritated by the trouser legs – 'There is no reason why Supermarine could not have included retractable undercarriage and flaps in their original design.'[5]

The Air Ministry even looked into the possibility of buying the PZ24 fighter from Poland, a gull-winged metal monoplane that was 4 mph faster than the F7/30 prototypes and could climb to 15,000 feet (4,572 metres) two and a half minutes faster. The bizarre prospect of Polish fighters defending Britain was avoided by the Air Ministry's conclusion that the PZ24 was insufficiently advanced. There were also the ethical considerations of buying a Polish plane to meet British fighter requirements when the country had a perfectly good aero industry of its own.

Mitchell responded with a series of alterations to his Type 224 that turned his ugly duckling into a swan. Off came the fixed trouser legs and on went a retractable undercarriage, the legs folding outwards.* The cockpit was enclosed and the wings were completely revised, losing 6 feet (1.8 metres) in span, gaining flaps and becoming thinner. There was a new tailplane and elevator, and four guns were placed in the wings outside the propeller diameter. This was more like it: the new Type 300 was said to be good for 265 mph. The Air Ministry, so often portrayed in Spitfire historiography as dismissive of this design, was encouraging. Air Commodore Cave-Brown-Cave praised the Type 300 to Dowding and a £10,000 contract was issued. Now Mitchell pressed on with improvements, adopting an advanced stressed-skin† construction in which the metal surface of the aircraft contributed to its strength, in contrast to the weak fabric skin of previous designs. This involved the use of sunken rivets, an expensive and time-consuming technique but which gave a smooth finish.

Despite reducing the wingspan and wing thickness even further, there was insufficient power, as the steam-cooled Rolls-Royce Goshawk engine only developed 660 hp. But now Mitchell, after a run of personal and professional misfortune, had a stroke of good luck. After the success of his Schneider Trophy aeroplanes he had kept on friendly terms with Rolls-Royce – indeed he still owned the car they had given him. They informed him that their new PV12 engine was already giving 800 hp and was expected to reach over 1,000 hp soon. He calculated that with this sort of power his new fighter could exceed 300 mph, maybe more with further improvements.

---

* This made the wheel track rather narrow, which resulted in the Spitfire being somewhat unstable on rough grass airstrips. One pilot called it 'a lady in the air but a bitch on the ground'.

† It should be pointed out that the 1913 Schneider Trophy-winning French Deperdussin monoplane had this construction, in plywood.

'The Spitfire had a German wing.' So claimed Dr John Ackroyd, fellow of the Aeronautical Society, in a recent interview.[6] This heresy may well be true. Mitchell's Canadian aerodynamicist Beverley Shenstone has long been credited with the Spitfire's distinctive wing design. As a student he realised that metal mono-planes were the future and that the Germans were now leading the field. After help from the British Air Attaché in Berlin he landed a job at Junkers in Dessau and met the aerodynamicist Ludwig Prandtl, who had drawn a double-elliptical wing plan, the two halves of which were composed of two different ellipses. Returning to England, Shenstone ended up working with Mitchell on the new fighter.

Mitchell and his team decided to redesign the wings. They dispensed with the cranked shape, necessary to keep the old fixed-undercarriage legs short (as with the Junkers Ju 87 Stuka). Now they decided to go with Shenstone's new elliptical design, which became the defining shape of the Spitfire. An ellipse gives the least induced drag for the maximum lift. It also gives a wide chord; the distance between the leading and trailing edge at the wing root where the wing joins the fuselage, and the wing tapers off more gradually towards the tip. This gives plenty of room for undercarriage, guns, and all the other paraphernalia that wings are expected to carry. In contrast, the chord of the tapered wing such as the wing of the Type 224 starts to reduce as soon as it moves away from the wing root.

There was also a subtle twist in the elliptical wings that was one of the secrets of the Spitfire's handling. The inner wing was set at a higher angle of attack (angle of the wing to the relative wind) than the outer wing. The inner part of the wing thus generated propor-tionately more lift, whereas the tip, at a smaller angle of attack, generated less. It meant that any stalling of the airflow began at the wing roots rather than the tips. The buffeting would give the pilot advance warning in tight turns, and according to test pilot Jeffrey Quill this saved many lives:

> The Spitfire's extremely docile behaviour in the stall was one of its greatest features. You could pull it well beyond its buffet boundary and drag it round with full power and little airspeed; it would shudder and shake and rock you from side to side, but if you handled it properly it would never get away from you. Whether they knew it or not, there are many pilots alive today who owe their survival to this remarkable quality in the Spitfire – and I am one of them.

Mitchell, whose illness had not improved his mood, discussed the wing shape with Beverley Shenstone and said: 'I don't give a bugger whether it's elliptical or not, just so long as it covers the guns.'

Was Beverley Shenstone a spy? The involvement of the Air Attaché suggests that industrial espionage might have been at play. Or was the Spitfire's wing just another consequence of the endless cross-fertilisation of the 1930s air industry?

Ellipses are generally held to be appealing, and the Spitfire was a symphony of ellipses. They are also hard to draw, involving two focal points. Straight lines and circles are easier to draw.

When the thirteenth-century Pope Boniface VIII was looking for an artist to work on the frescoes in St Peter's Basilica he sent a messenger to the painter Giotto asking for a drawing as a demonstration of his skill. Giotto took a brush loaded with red paint and drew a perfect circle without using a pair of compasses and handed it to the man. The envoy was angry, having expected a drawing of angels. He took the drawing back to the Pope, who understood the significance of the red circle, and Giotto got the commission.

Supermarine got the commission too – eventually. Installation of the new Merlin engine proceeded apace, and a number of improvements were made. The first was an extraordinary effect discovered by Frederick Meredith, a young scientist working at the Royal Aircraft Establishment at Farnborough. Thinking about how to use some of the waste heat pouring out of the aircraft engine, he experimented with placing the coolant radiator in a shaped duct which had a fast flow of air through it. Air flowing into this duct met drag resistance from the radiator surface and was compressed. As the air flowed through the radiator it was heated, raising its temperature and increasing its volume. The hot, pressurised air then exited through the exhaust duct which narrowed towards the rear. This accelerated the air backwards and the reaction of this acceleration against the Spitfire provided a small forward thrust, in fact enough to compensate for around 90 per cent of the drag of the radiator duct.[7] He had turned the radiator into a rudimentary jet engine. Later there were even plans to equip the Spitfire with an afterburner, by injecting fuel into the exhaust duct after the radiator and igniting it!

Meredith wrote in his paper that the thrust developed by the system would be increased if exhaust gases were also vented in the ducting downstream of the radiator. This would have involved convoluted exhaust pipes that would have actually sapped power, but there was a simpler solution. Ejector exhausts had already been thought of, and in September 1938 the Spitfire prototype, K5054, was fitted with these exhaust stubs, which are the only visible part of the engine when the aircraft is in flight.

The Rolls-Royce Merlin inhaled a vast volume of air at full power (about the volume of a single-decker bus per minute), and with the exhaust gases blasting out at 1,300 mph (2,100 km/h) it was realised that useful jet thrust could be obtained by angling the gases backwards instead of venting sideways. This thrust turned out to be the equivalent of 70 hp at 300 mph, or enough to increase the speed of the Spitfire by 10 mph (16 km/h). The first versions of these ejector exhausts had simple round outlets, while subsequent versions of the system used 'fishtail'-style outlets which slightly increased the thrust and reduced exhaust glare during night flying.

It only remained to name the aircraft, which was originally to be called the Shrew or Snipe. An alliterative 'S' was desirable, to go with 'Supermarine'. Sir Robert McLean, chairman of Vickers Aviation, owners of Supermarine in the 1930s, came up with the name Spitfire for the earlier Type 224 aircraft as he was in the habit of calling his elder daughter Annie 'a right little spitfire'. The word was used by Elizabethans to mean a ferocious woman, and McLean might have been thinking of Shakespeare's King Lear, who also had problematic daughters, and who played on the word thus:

'Rumble thy bellyful! Spit, fire! spout, rain! Nor rain, wind, thunder, fire are my daughters.'

On Christmas Eve 1936, the year in which the Spitfire prototype flew for the first time at Eastleigh, Hampshire, Annie was married. She lived until she was 100. The Air Ministry had reservations about the name, as did Mitchell, who preferred Shrew, but the boss prevailed, and Spitfire it was. Mitchell was unimpressed with the name – 'Just the sort of bloody silly name they would think of,' he said.[8]

The prototype Spitfire had been built in a tarpaulin-screened-off corner of the Supermarine works at Woolston, Southampton. They were worried about the curious eyes of German Lufthansa pilots, who regularly landed their seaplanes on the River Itchen nearby to have their airmail cleared at British customs. The first time the Merlin engine was run in a Spitfire was here.

The first engine run was carried out at night for security reasons, with the tail skid lashed to a holding-down ring on the quay normally used for tethering down flying boats ... the flames from the exhaust were spectacular, as was the noise. After the runs were completed, the wooden fixed-pitch airscrew and the wings were removed and the aircraft loaded onto lorries for the journey to Eastleigh aerodrome and re-erected there in our hangar, where, incidentally, the S6B was stored, less engine.[9]

Now carrying the official registration of K5054, the Spitfire prototype was first flown on 5 March 1936 by the Vickers chief test pilot Joseph 'Mutt' Summers. 'The new fighter fairly leapt off and climbed away,' reported Supermarine pilot Jeffrey Quill, who was standing next to Mitchell to witness the event. In his report Summers was full of praise: 'The handling qualities of this machine are remarkably good,' he wrote. Quill took over at the end of March 1936 and became the leading Spitfire pilot. His first flight was memorable for him: 'at once I felt good in that cockpit. I primed the Merlin carefully and it started first time. I began taxiing out of the north-east end of the airfield which was, of course, entirely of grass ...'

Quill took off and began to enjoy the flight:

the aircraft began to slip along as if on skates with the speed mounting up steadily and an immediate impression of effortless performance was accentuated by the low revs of the propeller at that low altitude. The aeroplane just seemed to chunter along at outstandingly higher cruising speed than I had ever experienced before, with the engine turning over very easily, and in this respect it was somewhat reminiscent of my old Bentley cruising in top gear ... As I chopped the throttle on passing the boundary hedge the deceleration was hardly discernible and the aeroplane showed no desire to touch down – it evidently enjoyed flying – but finally it settled gently on three points and it wasn't until after the touch-down that the mild buffeting associated with the stalling of the wing became apparent. 'Here', I thought to myself, 'is a real lady.'

Shenstone's wing design had evidently worked, but Mitchell was a worried man. His new design was barely faster than the Hurricane – 335 mph versus 330 mph – and it was more expensive and far harder to construct. He knew that unless the Spitfire was considerably faster than the Hurricane it was unlikely to be put into production. So he tweaked the design of the propeller and commissioned a Rolls-Royce contractor to apply an ultra-smooth high-gloss finish in a duck-egg blue. Quill flew the aircraft and found he could now reach 348 mph in level flight and 380 mph in a dive. Mitchell was now satisfied enough to release the prototype for the official Air Ministry tests.

These tests went well, and large orders were placed for the Spitfire. However, Supermarine were unable to build the aircraft fast enough, with the advanced construction causing delays. These delays were helpful to Rolls-Royce while they struggled to get the Merlin engine right. One of the most bizarre problems was the fact that some of the Spitfire wings built at the Woolston factory fitted the fuselages and some of them didn't. It took a clever detective to find the reason: the jigs (accurate frames) used to build the wings had been installed in a workshop erected on reclaimed land. Even though the workshops had been built on huge piles driven into the mud, the ground heaved up and down with the tide and the jigs also moved between high and low tide! As a result some wings had bends built into them that weren't supposed to be there.

* * *

The Spitfire had one lethal flaw. The fuel was contained in tanks just ahead of the pilot, and if ignited by enemy bullets flames would pour through the cockpit air vents and incinerate the pilot. Richard Hillary, a Battle of Britain Spitfire pilot, suffered exactly this fate, and unable to open a jammed cockpit canopy, suffered terrible burns to hands and face. He managed to bail out and became a member of the Guinea Pig Club, the group of burns patients treated by the pioneer plastic surgeon Archibald McIndoe. He persuaded the RAF to allow him to return to flying, and although he had difficulty handling a knife and fork he began to convert to light bombers. Both he and his navigator were killed when he crashed a Bristol Blenheim during a night training flight.

The Free French pilot Pierre Clostermann had this assessment of the Spitfire:

> For a pilot, every plane has its own personality, which always reflects that of its designers and colours the mentality of those who take it into action. The Spitfire, for instance, is typically British. Temperate, a perfect compromise of all the qualities required of a fighter, ideally suited to its task of defence. An essentially reasonable piece of machinery, conceived by cool, precise brains and built by conscientious hands. The Spitfire left such an imprint on those who flew it that when they changed to other types they found it very hard to get acclimatized.[10]

Mitchell had been diagnosed with cancer again in 1936, and had to give up work in early 1937. He used to watch the Spitfire test flights, though, and Jeffrey Quill would join him in his Rolls-Royce and tell him how things were going. Mitchell died in June 1937, aged just 42. One of his last ideas was for the Speed Spitfire, a modified version intended to make an attempt on the world landplane speed record of 352 mph which at the time was held by the millionaire Howard Hughes flying a racing aircraft. The Speed Spitfire would use a specially uprated Merlin engine.

This was a special sprint version of the Merlin, taken from stock and fitted with strengthened connecting rods and pistons. The pins that joined these, the gudgeon pins, are amongst the most highly stressed components in a piston engine, and these were reinforced, too. That was all that was necessary to make the sprint Merlin reliable. To make it powerful the supercharger pressure was raised to a staggering 27 lb per inch, when 6¼ lb per inch was used on the first production Merlins. Also, a special racing fuel was used. This was a witches' brew of 60 per cent benzol, 20 per cent methanol, 20 per cent petrol, together with an unhealthy dash of 4 cm^3 of Thomas Midgley's finest tetraethyl lead. The result was astounding: instead of the standard Merlin II's power of 1,310 hp, this engine more than doubled that output at 2,160 hp, delivered at 3,200 rpm. The all-important specific power-to-weight ratio was 0.62 lb/hp (compare this with the Wrights' 15 lb/hp). This showed more than anything else the potential that lay in the Rolls-Royce Merlin: the basic structure of this engine could withstand more than double the power. Lesser engines might throw rods, overheat and blow up, but at last the Merlin was reliable.

This special engine drove a coarse-pitch Watts wooden propeller of 10 feet diameter. A coarse pitch was needed to screw through the high-speed airstream at over 400 mph (643 km/h). Cooling this special Merlin necessitated fitting large coolant and oil radiators. The wingspan was reduced to 33 feet 9 inches (10.28 metres) and the wingtips were modified. All the gaps between the alloy panels were filled in and smoothed over, all remaining dome-headed rivets were replaced by flush rivets, and a more sloping racing windscreen was fitted. Lastly, the original tail skid of the prototype was reinstated, replacing the drag-inducing tailwheel.

After all this effort it was learned that the Germans had broken the world speed records twice in succession, first with a Heinkel He 100 V8 at 463.9 mph (746.6 km/h), then a month later with a Messerschmitt Me 209 V1 at 469.22 mph (755.14 km/h). It was decided to let the attempts with the Spitfire lapse.

It is fair to point out that the German planes were racing aircraft seeking propaganda victories, while the Spitfire was a production fighter aircraft. The Nazis loved speed and the propaganda victories it brought. The state-funded 'Silver Arrows', the Auto Union and Mercedes-Benz racing cars dominated Grand Prix racing until the outbreak of the Second World War in 1939.

The efforts with the Speed Spitfire did not go to waste, though. Supermarine gained valuable data on how to increase the Spitfire's maximum speed and Rolls-Royce proved the reliability, strength and power potential of the Merlin. All they now had to do was to get on with making it.

There was no point having the most reliable and powerful aero engine in the world if it couldn't be made in large numbers. Each Merlin required the manufacturing capacity of 40 car engines, and essentially they were hand-made by Rolls-Royce craftsmen. Shadow factories had to be built both to increase the number of Merlin engines built and also to disperse production so that the expected bomber offensive couldn't wipe out British aero-engine building at one stroke. In this new kind of war the machine tools of Germany would be fighting the machine tools of Britain.

Ernest Hives was more than willing to cooperate, but he warned the Air Ministry that it wasn't just a case of replicating the Derby factory:

The Rolls-Royce factory at Derby had been built and developed around the problem of producing high-class engineering in relatively small quantities with the capacity to change or modify the product quickly. In other words, I should describe the Derby factory as a huge development factory rather than a manufacturing plant. The very nature of the organisation necessitates a large proportion of skilled men who fortunately are available.

The Derby factory built all the Merlins that fought the Battle of Britain, and total Merlin production at Derby was 32,377. The first shadow factory for Rolls-Royce was set up at Crewe in 1938. Hives pointed out the difference: 'The production at Crewe has been planned on very different lines to Derby. We are making use of very much more unskilled and semi-skilled labour. Crewe cannot absorb modifications and alterations like the Derby plant.'

Crewe built 26,065 Merlins. The second shadow factory was at Hillington in Glasgow, which built 23,647 Merlins:

The chief advantage in going to Glasgow is that labour should be available. It should also be available for whatever size it is decided to make the factory. We have definitely decided against recommending any further extensions at Crewe because there is insufficient labour to draw upon, and, so far, insufficient houses.

Rolls-Royce was to experience huge delays and frustration in bringing the Crewe and Glasgow workforce up to the required standard. In Glasgow experienced foundry workers could not be found for love nor money, so specially trained women were employed to operate mechanised pattern equipment. At both sites the local authorities were incompetent at providing decent housing. Crewe Council had promised to build 1,000 houses by the end of 1938, but by February 1939 they had only issued one contract for 100. In May Hives became so angry he threatened to move the whole plant altogether. The situation was even worse in Glasgow, where building of the factory went ahead while the building of houses for the labour force stopped altogether.

The Ford factory in Trafford Park, Manchester was able to accept a contract for Merlins and tooled up for mass production of the engine. However, like Packard they complained that they were having problems with the Rolls-Royce tolerances: they were too large! The Rolls-Royce fitters were craftsmen used to selecting and hand-adjusting the large clearances to make a perfect fit. Ford had to redraft the Rolls-Royce technical drawings, but the result was cheaper engines of the same high quality, and by 1946 the Trafford Park factory had turned out 30,428 Merlins built by around 22,000 workers. The great engineer Stanley Hooker commented: 'once the great Ford factory at Manchester started production, Merlins came out like shelling peas at the rate of 400 per week. And very good engines they were too, yet never have I seen mention of this massive contribution which the British Ford company made to the build-up of our air forces.'*

The Ford Merlin plant was where the huge Trafford shopping centre is now located, and as author Carles Boix points out,[11] this neatly illustrates the great movements of destructive capitalism. Nineteenth-century Manchester was at the heart of the Industrial Revolution, when immensely wealthy factory owners employed low-skilled workers at low wages, resulting in huge inequality with no representation: Manchester with a population of 126,000 had no Members of Parliament, while Old Sarum in Wiltshire, with

* Hooker was a brilliant mathematician who by analysing the Merlin's super-charger increased the power by 30 per cent. It was this supercharger expertise that gave Rolls-Royce the lead in gas-turbine technology.

one voter, elected two MPs. Only 7 per cent of the male population had a vote. In August 1819 there was a rally of 60,000 pro-democracy reformers which was attacked by armed cavalry, resulting in 15 deaths and over 600 injuries. This became known as the Peterloo Massacre, and it could have been the beginning of a British revolution.

The next great stage of capitalism emerged in Detroit, where the use of modern assembly lines by factory owners such as Henry Ford demanded higher-skilled blue-collar workers who were better paid. Boix maintains that the growing wages, declining inequality and the rise of a middle class led to the flourishing of a benign form of democratic capitalism. This was the Detroit of the 1903 Packard Automotive Plant, where the Rolls-Royce Merlin was built. This building now lies in ruins, the result of the next great stage of capitalism; the Information Revolution that began in Silicon Valley in the 1970s, which requires fewer, more highly educated workers at the expense of the blue-collar working classes. Jobs in Britain and the US have gone offshore, and once again inequality has risen sharply. The signs are that capitalism and democracy are now drifting apart.

Life was tough for the Rolls-Royce shadow-factory workers and their hours were punishing. They were working shifts of twelve hours a day for seven days a week, then twelve hours a night for six nights. As dawn approached and their energy flagged the night workers sang to keep their spirits up. A few months after the war began, with much time being spent crammed in the underground air-raid shelters, absenteeism started rising due to physical and mental fatigue. In a bid to resolve the problem the working hours were *reduced* to 82 a week, and one half-Sunday a month was granted to the workforce (48 hours a week is the present UK legal maximum, together with 28 days annual holiday).

There was another problem. Even though Hitler was at the point of invading in 1940 there were strikes that threatened production. In Crewe there was a strike when women were assigned to capstan lathes that had previously been operated by men. The union claimed that skilled labour had been displaced. The situation was worse in Glasgow, as Hives fumed: 'The district is seething with communists, and strikes and threats of strikes occur the whole time.'

His Glasgow factory manager, Bill Miller was more analytical:

> ... as nearly all fit males had already been called up for war service, only those men who were unfit for the army, or were conscientious objectors, were available for employment. Unfortunately, a large number of the conscientious objectors were Communists and many of them managed, by hook or by crook, to become shop-stewards – with militant attitudes. Many of the shop steward committees were inclined to encourage strike action for the most trivial reasons.

After the collapse of Norway and the resignation of Chamberlain, Winston Churchill formed an all-party coalition government. Of the five members of his War Cabinet formed in May 1940 two, Clement Attlee and Arthur Greenwood, were Labour MPs. Other Labour members held ministerial positions, the most important of whom in terms of labour relations was Ernest Bevin, Minister of Labour and National Service. Clearly Bevin's standing as leader of Britain's biggest trade union (the Transport and General Workers Union) was intended to deflect the kind of opposition to the control of labour which had been a problem during the First World War. Largely due to these appointments the unions cooperated with the government, helping to organise production and setting targets.

Communists in Britain had little commitment to the war effort until the Soviet Union entered the conflict in 1941. They refused to be bound by the Order 1305 ban on industrial action and during the first few months of the war there were over 900 strikes. Bevin was anxious to avoid the labour unrest of the First World War and tried hard to encourage conciliation and avoid conflict and prosecution.

The Second World War came home to British citizens in a way the First World War had not; Guernica had provided a terrible example and the government prepared for saturation bombing of civilians. Gas attacks were expected, and by the outbreak of war 38 million masks had been issued, with special ones for children and babies. Children were evacuated from cities into the country in Operation Pied Piper. Women had a double burden, as in addition to their usual role in the family they were conscripted to work in wartime industries such as the production of the Rolls-Royce Merlin. Before long women made up one-third of the workforce.

During the First World War two female workers had been dismissed at a factory in Newcastle for wearing trousers outside the factory gates, prompting the 17 other women working at the firm to go on strike and also to be dismissed.[12] But during the Second World War either trousers or the siren suit became fashionable, so called because it could be pulled on quickly during an air-raid warning. Churchill had invented this one-piece boiler suit in the 1930s as his personal leisure wear and was often filmed wearing his pin-striped version. Bizarrely, he even wore his when meeting US President Franklin D. Roosevelt, General Dwight Eisenhower and Soviet leader Joseph Stalin. He can therefore be held responsible for 'onesies'.

Film of women working on assembling the Merlin shows them wearing the Victory Roll hairstyle which featured large curls of hair framing the face. The name was a reference to the fighter plane manoeuvre. They also wore red lipstick, and most of them smoked cigarettes on the production line. At the Derby Nightingale Road factory women were allowed to leave their shift three minutes early in order to get home and prepare food for the men.

Women pilots of Air Transport Auxiliary flew the Rolls-Royce Merlin when ferrying aircraft around Britain between factories, maintenance units and airfields. Pilots of the ATA were considered unsuitable for combat flying due to disability, age or gender. There were old, short-sighted, one-eyed, one-armed and one-legged pilots (not all at once), humorously referred to as the Ancient and Tattered Airmen: ATA. This was incorrect, as 166 of them were women, who attracted enormous attention in the popular press. These women received the same pay as men of equal rank in the ATA, and that was a milestone, as it was the first time that the British government gave equal pay for equal work.

It helped to be wealthy and upper-class. Diana Barnato had been a debutante and was presented to King Edward. Her grandfather had been the immensely wealthy Barney Barnato of the De Beers mining company, and her father was Woolf Barnato, Chairman of Bentley Motors. She became fascinated by aeroplanes when young, and got her pilot's licence at the age of 20 at the Brooklands Flying Club in Surrey. In 1939 she was given a Bentley motor car for her 21st birthday. She applied to the ATA in 1941, was accepted and eventually delivered 260 Spitfires and hundreds of Hurricanes, Mustangs and Tempests. With more training she converted to twin-engined types such as Mosquitos, Blenheims and Wellingtons, usually flying solo.

Barnato had a tragic personal life. She was engaged to a Spitfire pilot who was killed in a drunken flying accident. He was Squadron Leader Humphrey Gilbert, who attempted to fly home after drinking several bottles of Benskins Colne Springs, a 9 per cent alcohol beer. Drinking with him was Flight Lieutenant David Gordon Ross. When it was time to go home the pair tried to borrow a two-seater Magister aircraft. When the flight sergeant realised the officers were drunk he claimed that the aircraft was unserviceable, so they took a single-seater Spitfire instead, with Ross sitting in Gilbert's lap. Control of the aircraft was lost not long after taking off from Great Sampford, and the fighter spun in, killing both occupants.[13]

Barnato then married a Wing Commander, who was killed six months later while flying a Mustang fighter in bad weather. Diana promised herself never to marry again, and she became the mistress of a married man, who was another pilot and racing driver. By him she had a son.

After the war she established the world air speed record for women after persuading the Air Minister to let her fly an English Electric Lightning T4 to Mach 1.6 (1,262 mph, 2,031 km/h). She was the first British woman to break the sound barrier.

As if piston aero engines weren't complicated enough, the war was about to make them more complicated still. As we have seen, during the First World War propellers (or airscrews) were carved out of wood and had a fixed pitch. This was a compromise: during take-off when the air speed was low the pitch needed to be fine, in other words if the propeller really was an airscrew it would screw itself forward at slow speed. At high air speeds the propeller (spinning at the same revolutions) needed to be at a coarser pitch so that it could keep up with the faster air and therefore screw forward at higher speed. A propeller stuck with a fixed pitch was always going to be an unhappy compromise; what was needed was a propeller that would smoothly change pitch as the air speed increased. Even in 1917 the Royal Aircraft Factory was experimenting with variable-pitch propellers. By 1939, nine out of ten

Mercury and Pegasus engines made by Bristol were two-position propellers made by a competitor, de Havilland.

A word about the Bristol Aeroplane Company: unlike their competitor Rolls-Royce, they made their own aircraft as well as their aero engines. The first, the Bristol Boxkite, was demonstrated in November 1910 on Durdham Down at Bristol, a stone's throw from where this book is being written. We have already seen the success of their FE2 fighter during the First World War, and their Bristol Bulldog was the mainstay of the RAF between 1930 and 1937. They had concentrated on air-cooled radial engines during the Thirties and their enormous 39-litre, 14-cylindered two-row sleeve-valved Hercules engine always shaded the Merlin engine in power output: the 1939 Hercules I engine gave 1,290 hp, which soon improved to 1,375 hp in the Hercules II. The main version was the Hercules VI, which delivered 1,650 hp, and the late-war Hercules XVII produced 1,735 hp. These were big engines with a large frontal area, so they were more suitable for bombers or large fighters like their Bristol Beaufighter. Many British front-line aircraft were powered by Bristol air-cooled radial engines.

After the war it was possible to drive a Bristol 400 car from the works at Filton to the aerodrome at Lympne Airport in Kent, board a Bristol Freighter powered by Bristol Hercules 734 engines and fly to Le Touquet on the northern coast of France. Later Bristol was involved in the supersonic transport studies which culminated in Concorde, which was powered by Bristol Olympus jet engines.

Before the war Roy (later Sir Roy) Fedden, chief engineer of the engine department at Bristol, was concerned about the slowness of development of their bought-in propellers. He decided to approach his main competitor Rolls-Royce and suggested a joint venture to consolidate both companies' propeller development efforts and make a better design. This was agreed, the Air Ministry approved, and Rotol Airscrews Ltd was born on 13 May 1937. The name, a

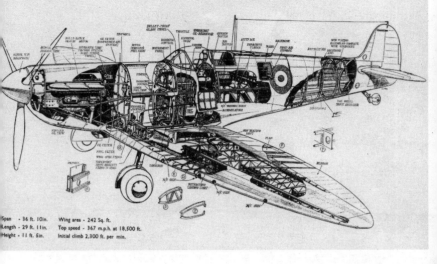

THE SUPERMARINE SPITFIRE II SINGLE-SEAT FIGHTER
(1,030 h.p. Rolls-Royce Merlin III motor)

Span - 36 ft. 10in.   Wing area - 242 Sq. ft.
Length - 29 ft. 11in.   Top speed - 367 m.p.h. at 18,500 ft.
Height - 11 ft. 5in.   Initial climb 2,300 ft. per min.

The Supermarine Spitfire Mk II.

contraction of ROlls and BrisTOL, was suggested by the wife of Bill Stammers, the first general manager. The result was that variable-pitch propellers were ready for the Hurricanes and Spitfires of the Battle of Britain, and according to Air Chief Marshal Tedder this was one of the three deciding factors in the conflict, the others being radar and 100-octane fuel. By the end of the war Rotol were building five-bladed propellers for late-model Spitfires.

But back at the beginning of the war Rolls-Royce had got the bit between their teeth and started to do what they were really good at: making their best engine even better. Afterwards Cyril Lovesey, Chief of Merlin engine development, revealed what Rolls-Royce's secret was: changes were infiltrated into production in small digestible amounts that did not stop the production lines. The fitters were skilled enough to accommodate design changes that would stop an ordinary mass-production line. This was all part of the genius of Rolls-Royce.

CHAPTER FIFTEEN

*It is perhaps as difficult to write a
good life as to live one*

To paraphrase Lytton Strachey; the history of the Second World War will never be written: we know too much about it. There is such a vast accumulation of material left to us that no human historian could sift through all of it and arrive at any kind of coherent conclusion. Instead, we might take Strachey's advice and practise a selective ignorance. In *Eminent Victorians*, his account of the Victorian Age, he tells us:

> ignorance is the first requisite of the historian – ignorance, which simplifies and clarifies, which selects and omits, with a placid perfection unattainable by the highest art … He will row out over that great ocean of material, and lower down into it, here and there, a little bucket, which will bring up to the light of day some characteristic specimen, from those far depths, to be examined with a careful curiosity.[1]

Mass-production of Merlin engines at a Rolls-Royce factory 'somewhere in Britain' in 1942. Here crankcases are readied for the next stage.

We have lowered our bucket and come up with the Rolls-Royce Merlin. And one way to view the Second World War is by following the story of this engine during the conflict. And if the reader hears any noises offstage during the rest of this story, don't worry, it's just the war grinding on.

Let's look at our specimen more closely. We have fished out a Merlin III, so this is an early variant that fought in the Battle of Britain, made in the Derby factory and delivered in July 1938. There were at least 70 versions of the Merlin. Although the prototypes were given letters – PV12, B, C, D, E, F and G – the production engines had Roman numerals until they got to XX, when perhaps for intelligibility reasons Rolls-Royce shifted to the Arabic numerals 21 to 266 for the wartime versions.

In appearance this Merlin engine is a large, black-painted lump of metal about the size of an office desk, made up of around 14,000 components. We know it can produce 1,440 horsepower at 3,000 rpm, but let's see how this forest of gears, shafts and lumps of metal actually works.

As we have seen, it has 12 **cylinders** arranged in two banks of six, which are mounted on an alloy crankcase at an included angle of 60°. The banks are denoted A for starboard and B for port. They are cooled by a mixture of glycol and water, so there are no cooling fins visible. In this Merlin III the cylinder blocks and heads are still cast together in a monobloc arrangement, but we know that later versions separated these. The cylinder liners are wet; that is, they are in direct contact with the coolant, and so the exterior of each liner is nickel-plated to resist corrosion. There is a seal recessed into a circular groove around the combustion chamber that mates with the top of the liner, and this has to keep in the high-pressure gases of combustion and keep out the boiling coolant. It is this seal that caused problematic leaks. The liners are almost totally sunk into the block, a small protrusion locating them in the spigoted circular holes in the crankcase.

At this point spare a thought for the poor fitters. When a leaky Merlin came in, boiling coolant spurting everywhere, they had to replace these seals. The same procedure had to be followed as on the winning Supermarine S6B in 1931: the combined cylinder block and head had to be lifted off the 14 studs and six pistons, the seals replaced, then the whole heavy lump carefully slid down over the pistons, immense care being taken not to break the brittle piston rings.

The **crankcase** is the long shiny casting at the bottom that contains and supports the crankshaft. It is made of RR50, an aluminium-nickel alloy developed by Rolls-Royce. The forward part of the crankcase forms the rear half of the propeller reduction-gear casing. It has 12 large circular holes bored in it to take the bottom of the liners, and a forest of 28 long studs to hold down the cylinder blocks. Inside the crankcase are partitions or internal webs which are bored to support the crankshaft. Each bearing journal takes plain half-shell lead-bronze bearings which support the crankshaft, which spins at up to 3,000 revolutions per minute, or 50 times a second. The centre bearing has to support a load of 14 tons at full power.

The bottom half of the crankcase looks rather like the sump of your car engine, if you peer underneath. It is an alloy bathtub that is usually filled with hot oil draining from the engine. It has two oil pumps in it; a scavenge pump to retrieve the oil and pump it to the oil tank, and a pressure pump to force the cooled oil back around the engine. The oil tank contains enough oil to lubricate the engine for a short time when it is flying upside down or experiencing negative g. Two brass plates on the left or port side inform us of the engine data and the cylinder firing order, so we know which Merlin this is.

The **crankshaft** is the heart of the engine. The Merlin crankshaft began life as a 500 lb chrome moly steel forging, and when it was finished it weighed just 120 lb. Every ounce had to be machined away by hand on manually operated lathes and mills in 140 separate operations. The machinists were very often women. It had been machined all over, heated up and then nitrogen-hardened. It has six balanced throws for six pairs of connecting rods. Car mechanics would find it strange that the crankpins are hollow and capped, but they would be familiar with the fact that the crankshaft is drilled throughout its length. This allows the passage of high-pressure oil which lubricates the main bearings and the big-end bearings which are mounted on the crankpins. At the front the crankshaft drives the propeller reduction gear and at the rear it drives the supercharger.

The steel **connecting rods** would be strange to a car mechanic, too, but perhaps not to an historically minded marine engineer. They are arranged in pairs; one of them a forked rod and one a plain rod. The plain rod is assigned to cylinder bank A, and it runs on a bearing contained in the forked rod, assigned to bank B. This bearing in turn runs on a crankshaft crankpin. The top ends of the rods carry the **gudgeon pins** in a bronze bush. These pins connect the rods to the pistons as wrists connect your arms to your hands.

The **pistons** are forged from another Rolls-Royce aluminium alloy: RR59. There are five grooves machined in each one containing, from the top: three compression rings to seal in the gases of combustion. These are made of hard-wearing cast iron. Below these is a spring-steel oil-scraper ring, which has to scrape excess oil from the cylinder bore. It has holes in it to allow the oil to flow back inside the piston. Finally, beneath the gudgeon pin there is another oil-scraper ring. When desert sand wore out the Merlin engines in the campaign against Rommel, it was these rings that wore out and let oil up into the combustion chambers, resulting in blue smoke, which is the sure sign of a worn-out piston engine the world over.

The **combustion chambers** are formed inside the one-piece cylinder blocks and cylinder heads. They contain two inlet **valves** and two exhaust valves. These are forged from austenitic nickel-chrome steel: what we would call stainless steel. As we have seen, these started off in earlier engines looking like a penny on a stick. As they glowed red-hot, then white-hot, they stretched into a tulip shape, and so RAF engineers in a fit of empathy suggested that as they were trying to take that shape perhaps they should be allowed to start life that way. So the Merlin valves are of the tulip, or trumpet shape. As we have seen, the exhaust valves had hollow stems into which sodium was inserted to enable cooling, then a cap was welded on to seal the chamber. All the valves have ends made of Stellite, a cobalt-chromium alloy designed for wear resistance. Soon this metal would be important in gas turbines made by Rolls-Royce. Later the exhaust-valve seats were also made of Stellite. The valves have double valve springs to close them. There are two **sparking plugs** in each cylinder, and cleverly, these are screwed into aluminium-bronze inserts with a left-hand thread, so they don't screw out of the head when the sparking plugs are removed.* The manufacturers told the general public that the insulators of the sparking plugs that kept their fighters in the sky were made of the same stuff as rubies. In this they were right, in the same way that coal is related to diamonds, for the stuff was merely sintered aluminium oxide.

---

* Left-hand threads have foxed many an amateur mechanic. Instead of unscrewing anticlockwise they have to be unscrewed clockwise. Owners of post-war Rolls-Royce cars trying to remove the wheels on the left side of their cars may not have known that Rolls-Royce put left-hand threaded wheel nuts on this side. This was to avoid wheel rotation loosening the nut. This has led to much cursing in tyre-fitting bays.

There is a single **camshaft** mounted above each bank of cylinders in seven pedestals, running at half crankshaft speed for reasons of valve timing. The lobes on the camshafts open the valves through levers called rockers, which have adjustable tappets at the ends to maintain the correct valve clearance. A modern car mechanic might be puzzled by the way the camshafts are driven; instead of a toothed belt or chain drive there is a bevel gear at the rear of each camshaft. This is driven by a vertical shaft from the wheelcase, driven from the crankshaft at half speed.

The **wheelcase** is a tangled nest of gears, all driving in different directions. There are drives for the two magnetos on either side of the case, driven from a single upright shaft (this caused trouble, as we have seen), drives for the electrical generator, the coolant pump, the electric starter and manual turning gear. Yes, you read that right: the **manual turning gear**. These 27-litre engines were issued with a starting handle. Anyone who has attempted to hand-start a 1-litre car may have been shocked at (a) how hard it is to turn an engine over against compression and (b) how painful it is when the engine backfires and breaks your wrist. It must have taken a superman to hand-start a Merlin.*

We have seen how the **exhausts** were angled backwards to give a jet-ejector effect, and the **induction manifolds** are fed with compressed air and petrol mixture from the **supercharger**, which on this early model of Merlin is a single-speed, single-stage version. A forged-alloy 10¼-inch-diameter impeller is driven from the

---

* Herein lies a curious story. When Peugeot named a new model in the 1920s, they wanted to call it '21' after the horsepower, like the Rolls-Royce 40/50. The metal 2 and 1 digits of the '21' model were attached either side of the large hole for the starting handle, which inadvertently created the number '201', which Peugeot finally chose as the model name. All Peugeot model names sprang from this middle zero convention and remain to this day as the Peugeot hallmark. When Porsche wanted to call a new car 901 Peugeot objected, and so Porsche had to adopt the number 911.

A Merlin sectioned cylinder head. Note the parallel valves and the hollow camshaft for lightness.

wheelcase by a spring-drive at over eight times the speed of the crankshaft. The impeller is enclosed by an aluminium casing that feeds the compressed mixture into the cylinders. The air/fuel mixture is prepared in the **carburettor**, which is an immensely complicated alloy box full of holes, drillings, butterfly valves and aneroid capsules (to correct for altitude). Like motor-car carburettors, these early Merlin S.U. carburettors had a float chamber rather like a lavatory cistern: fuel would flow into the chamber via a needle valve opened or closed by a brass float. When the fuel reached the correct level the float rose up in the chamber, closed the valve, and the fuel flow stopped. When the engine used up some fuel, more was supplied by the float dropping and opening the needle valve. So far, so good – on a car. Unlike cars, though, aircraft have a habit of flying upside down, or suddenly diving.

Under these negative-g conditions the fuel would rise to the top of the float chamber and the float would rise, cutting off the fuel flow. The engine would suffer a weak-mixture cut-out and momentarily lose all power. When the aircraft came out of the dive or roll, and positive g was experienced again, the float chamber would now be empty and the float would open the valve too much and the engine would now suffer a rich-mixture cut-out.

During combat this proved to be a major problem for Hurricane or Spitfire pilots. If they dived in pursuit of a Messerschmitt Bf 109 the Merlin would cut out in negative-g conditions, then with a great jerk it would start running again when positive g was experienced. The Luftwaffe plane, having direct injection unaffected by negative g, could either escape his pursuer or reposition himself to shoot the RAF aircraft down.

Beatrice Shilling was an engineer at the Royal Aircraft Establishment at Farnborough who specialised in aircraft carburettors. She studied this problem and came up with a quick-fix interim solution. She discovered that the Merlin's fuel pump delivered nearly twice the fuel needed by the engine when delivering maximum power, so she invented a simple solution: a restrictor in the fuel line with a small hole drilled in it which would reduce the fuel flow. At first this looked like a brass thimble; later it became a washer. This solved the rich-mixture cut; then RAF pilots learned a wing-down entry to a dive; a quick half-roll which kept a positive-g element in the dive. This ameliorated the weak-mixture cut. She led a small team around the RAF squadrons in March 1941, installing the restrictors in their Merlin engines. This device was understandably popular with pilots and attracted a vulgar epithet, but a proper solution wasn't found until the introduction of the Bendix pressure carburettor in 1943. This injected fuel into the eye of the supercharger impeller and was really what we would call today throttle-body fuel injection.*

The **ignition system** comprises two **magnetos**, which are self-powering devices that provide sparks for the sparking plugs. They are still used in aviation piston engines, as they are regarded as more reliable than the automobile battery, coil and distributor system once found in cars. There is one magneto mounted on each side of the wheelcase, and each one has 12 cables that connect to the sparking plugs. For redundancy reasons the cables from the magneto on the starboard side connect to the sparking plugs on the inlet sides of the cylinders, and the port-side magneto connects to the exhaust sides. In this way if one magneto is shot to pieces the remaining one can keep the engine running. If both fail there is a deafening silence.

We have examined our specimen of the Merlin engine, now let's see it go to war.

We have learned from the bombing of Guernica that the war was probably going to be won or lost by the aeroplanes and aero engines fielded by the combatants. We have also seen how the Rolls-Royce company was still desperately trying to make enough Merlin engines in the late 1930s, and we know that fighter aircraft were not available in sufficient numbers for the British defences.

These facts shed light on the much-discussed actions of Britain's Prime Minister Neville Chamberlain, who is now mostly remembered for his foreign policy of appeasement. In particular, he was

---

* Beatrice Shilling OBE PhD MSc CEng bought a motorbike at the age of 14 and raced motorbikes at over 100 mph at Brooklands. It was said that she refused to marry her bomber-pilot husband until he too had been awarded the Brooklands Gold Star for doing the same. She was a fervent believer in women's rights and worked at the RAE on the Blue Streak nuclear missiles after the war.

excoriated for his signing of the Munich Agreement in September 1938, conceding the German-speaking Sudetenland region of Czechoslovakia to Germany. Critics forget that as Chancellor in 1934 it was Chamberlain who pushed an air force expansion scheme onto a reluctant Air Minister.

More recently, historians have pointed out that going to war with Germany in 1938 would have been disastrous for the UK, as the country was completely unprepared, and Chamberlain knew it. A glance at the forces arranged on either side makes for sobering reading: in September 1938 the Germans had 1,200 modern bombers, and all Fighter Command had to oppose them were 93 eight-gun fighters and 573 obsolete biplanes. There were no Spitfires ready and the Hurricanes, not yet having heating for their guns, could not use them above 15,000 feet (4,572 metres).

Chamberlain was well aware of these facts before the Munich negotiations. He later wrote to his sister: 'I stick to the view I have always held that Hitler missed the bus in September 1938. He could have dealt France and ourselves a terrible, perhaps a mortal, blow then. The opportunity will not recur.'[2]

Perhaps Chamberlain really did employ delaying tactics at Munich. Whatever his thoughts, he effectively handed the Germans Czechoslovakia – 'a far-away country … people of whom we know nothing'. Jan Masaryk, the Czech minister in London, was coldly furious:

'If you have sacrificed my nation to preserve the peace of the world I will be the first to applaud you. But if not, God help your soul.'

Had Neville Chamberlain outmanoeuvred Hitler in a Machiavellian move? He led Britain through the first eight months of the Second World War, when he was replaced by the more pugnacious Winston Churchill, a man better suited to war. Cometh the hour, cometh the man. Chamberlain, like our subjects Henry

Royce and Reginald Mitchell, also succumbed to bowel cancer. He died in November 1940. His last thoughts on his legacy were: 'without Munich the war would have been lost and the Empire destroyed in 1938.'[3]

He might have been surprised to learn that many historians today agree that the hostility of the US, which had grown into the leading superpower in the 1930s, would have soon led to the collapse of the British Empire. But Britain has never quite got over losing her empire, nor has she quite got over winning the war.

'There will be no European war this year or next year either.' This headline in the *Daily Express* appeared on 30 September 1938 to mark the Munich Agreement, and was reprinted in various forms at regular intervals well into 1939. In the opening sequence of Noël Coward's film *In Which We Serve*, the opening sequence includes a copy of the *Daily Express* floating in the dock jetsam bearing the headline 'No War This Year'.

It was probably the worst prediction in newspaper history, and haunted the paper's proprietor, Lord Beaverbrook, who promptly blacklisted Coward. Wishful thinking had overwhelmed his judgement, as he had supported the appeasement policies of the Chamberlain government. Beaverbrook was another one of Britain's egregious newspaper magnates whom Stanley Baldwin had bitterly described as exercising 'Power without responsibility – the prerogative of the harlot throughout the ages'.

Beaverbrook was given power *with* responsibility in May 1940 when Winston Churchill appointed him as Minister of Aircraft Production. His dictatorial manner made him no friends, but he drove aircraft and aero-engine production numbers up in the same way that he had driven newspaper circulation. He used the telephone as a weapon, urging manufacturers on to further efforts. When Beaverbrook rang Sir Charles Craven, chairman of Vickers, his secretary nervously informed Lord Beaverbrook that Sir Charles was on the lavatory attending to the wants of nature. When Beaverbrook demanded that he come to the telephone at once, Sir Charles shouted through the door: 'Tell his Lordship I can only deal with one shit at a time.'[4]

As well as badgering manufacturers, Beaverbrook unified the nation around his Spitfire Fund, using the power of the press. He raised £13 million, the purchase price of around 2,600 Spitfires. He claimed that each aircraft cost £5,000, a figure he thought the public would find achievable. In fact, they cost £12,604, or around £700,000 at the time of writing. This included the cost of a brand-new Merlin engine (£2,000, now £106,000) and propeller (£350, now £18,600). Compared with the astronomical cost of modern fighter aircraft, this seems like a bargain. The cost of one F-35 fighter jet is more than £78 million! One of the oddest donations came from 2,500 British prisoners of war held at Oflag VIB camp near Warburg, who contributed one month's pay 'to charity', the money being deposited with the German authorities who unwittingly forwarded it to England via the Swedish Red Cross. This paid for one Spitfire which was named *Unshackled Spirit*.

The Second World War began when Germany invaded Poland on 1 September 1939, and France, the United Kingdom, Australia and New Zealand declared war two days later. There followed a period of relative inaction on the Western Front known as the 'phoney war' (in France, *la drôle de guerre*), which ended when German forces invaded neutral Norway on 3 April 1940, seeking to secure their supplies of iron ore from neutral Sweden. This they achieved despite resistance from the Allies. The Germans next invaded the Low Countries and France on 10 May 1940. Preparations to defend France seemed lackadaisical: British Expeditionary Force officers helping to man the Maginot Line had so much time on their hands that they imported packs of foxhounds and beagles in 1939. They were thwarted by the French authorities in their attempts to introduce live foxes. When it was suggested that the Black Forest be bombed with incendiaries to burn German ammunition dumps, the Secretary of State for Air responded that the forest was 'private property' and could not be bombed; neither could weapons factories, as the Germans might do the same. In retrospect there seemed to be a certain lack of aggression on the part of the Allies, and the German generals could not believe their luck. At the Nuremberg Trials, General Siegfried Westphal stated that if the French had attacked in force in September 1939 the German Army 'could only have held out for one or two weeks'. And the Chief of the Operations Staff of the German Armed Forces High Command, Generaloberst Alfred Jodl, said that 'the only reason we did not collapse in 1939 was due to the fact that during the Polish campaign the 110 French and British divisions in the West were completely inactive against the 23 German divisions'.

The war began for Rolls-Royce Merlins in Scotland, when Spitfires from 602 and 603 Squadrons (from Glasgow and Edinburgh respectively) shot down two Junkers Ju 88s and a Heinkel He 111 over the Firth of Forth on 16 October 1939.* The German aircraft were obliged to operate from their own country's airfields, as they had not yet overrun France and the Low Countries. They were attacking Royal Navy ships at Rosyth on the Firth of Forth. The first German aircraft to be shot down on the British mainland was

* A Heinkel 111 bomber, wheels up, propellers bent, crash-landed on the runway of an RAF airfield. The pilot thought he had ditched in the sea and before the astonished eye of the airfield control officer a door was opened, a dinghy was thrown out and two of the crew – bootless for easy swimming – dived out onto dry land.

an He 111, with both 602 and 603 Squadrons claiming the victory.* The engines were immediately removed and taken away to be checked for fuel type, metallurgical qualities, design and power.

If any reader wonders why aero-engine makers seemed obsessed with horsepower, they might take a look at the tragic history of the Fairey Battle light bomber. On the day after war was declared ten squadrons of Battles were deployed to French airfields. Then, on 20 September 1939, a German Messerschmitt Bf 109 was shot down by a Battle gunner during a patrol near the Belgian border; and this event is recognised as being the RAF's first aerial victory of the war. Frankly, the Bf 109 pilot must have been asleep, because his aircraft far outclassed the Battle. So far, so good.

Then on 10 May 1940 the full might of Blitzkrieg smashed into northern France. On the same day, Winston Churchill became British Prime Minister. Unescorted Fairey Battles were sent to bomb the armoured columns, but with only two machine guns and too little power from the single Rolls-Royce Merlin I they were massacred by German Bf 109 fighters that were nearly 100 mph faster. Thirteen Battles were lost on that first day.

Vital bridges had not been mined in advance of the German armour, and on the second day nine Belgian Air Force Battles attacked bridges over the Albert Canal, losing six aircraft. In a later RAF sortie that day against a German column only one Battle out of eight survived.

On the third day of this nightmare, 12 May 1940, bridges over the Albert Canal were again the target, with four Battles shot down

---

* A Luftwaffe airman, Heinz Joseph, parachuted to the ground at Scoraig, North West Scotland, knocked at the window of a remote croft and asked for help. He spoke several languages except English, but a young daughter of the house, Joey Stewart, found that they could converse in Latin. The heel marks where he had landed heavily in the dark were still plainly visible in the peat just above the shore of Annat Bay until recently.

and the last one crash-landing. One span of the bridge was bombed successfully but the Germans swiftly replaced it with a pontoon bridge. And on 14 May all available Allied bombers were sent in, but were met by a huge force of German fighters: 35 Battles were lost this time, out of 63 sent in on the hopeless attack. It was a turkey shoot.

In just six weeks nearly 200 Fairey Battles were destroyed, with 99 lost between 10 and 16 May. Fifty per cent of Battles failed to return from sorties, an appalling rate of attrition. By the end of the year the aircraft had been withdrawn from front-line service. It was one of the most disastrous aircraft ever deployed by the RAF. The simple reasons were that it was too heavy, too slow, and the Merlin I underpowered. Low-level attack missions were given instead to single-engined fighter-bomber aircraft, such as the Merlin-engined Hawker Hurricane, and later the Napier Sabre-engined Hawker Typhoon. The Germans suffered similar attrition to their single-engined Ju 87 Stuka dive-bomber during the Battle of Britain. Something had been learned: horsepower was going to win the war in the air.

We last met Louis Strange in the First World War, hanging from the machine gun of his upside-down biplane. On 21 May 1940 during the Second World War this remarkable man was acting as Aerodrome Control Officer in Merville, France, having insisted on rejoining the RAF at the age of 49 (the flying duties age limit was 32). The German advance was nearly upon the airfield, and Strange was busy repairing some Hurricane fighters which had been too hastily sabotaged by the British to avoid them falling into German hands. This had been done by cutting the cables to the varia-ble-pitch propellers. Strange realised that some of these valuable Hurricanes could be made serviceable and flown back to England. He and his fitters found some telephone cable and reconnected the propellers, which were set into fine pitch for take-off. A hard tug on the cable would change into coarse pitch for high speed.

As they worked on the aircraft a ferocious dogfight between Hurricanes and Messerschmitts broke out above them. One of the Hurricanes was hit and fell towards the ground, the pilot bailing out and landing on the airstrip. The pilot was taking off his parachute when Louis Strange walked up to him and asked: 'Would you like another Hurricane?' Pilot Officer Tony Linney jumped in and headed for England. When 229 Squadron returned to RAF Digby they thought Linney was lost, but instead found that he had got home before them in a different aircraft.

Meanwhile, Louis had climbed into another of the repaired Hurricanes, a type he had never flown before. He managed to take off and yanked hard on the telephone cable. It worked. Shortly afterwards his unarmed aircraft was set upon by six Messerschmitt 109s, but by pretending to get into a firing position and then roaring up a village street and then down a chateau driveway he managed to shake them off. Eventually he got his Hurricane home to RAF Manston, fit to fight another day.

The German aero engines put up against the Rolls-Royce Merlin were formidable adversaries. Their manufacturers had been prevented from building any high-powered aero engines by the terms of the Versailles Treaty, but on 28 June 1926 representatives of Daimler-Motoren-Gesellschaft and Benz & Cie signed the agreement for the merger of the two oldest automobile manufacturers in the world. In 1927 they quietly made a V12 of 54 litres capacity, producing 1,030 hp, then they made a smaller one of 30 litres and then turned it upside down in 1931. When Hitler became Chancellor in 1933, they threw off the shackles of the treaty and produced their enlarged DB600 line of engines which kept Germany at the forefront of air power until her defeat in 1945.

Some might say that these Daimler-Benz engines were the best-designed of their era because they needed so little development to reach their potential. Daimler-Benz rarely made a design mistake that needed rectification. Like the 27-litre Rolls-Royce Merlin, the 34-litre DB600 had 12 cylinders liquid-cooled in a V configuration, but as we have seen the whole engine was inverted in a Λ shape. Like the Merlin the engine was supercharged with two inlet valves and two exhaust valves per cylinder opened by a single camshaft per bank of cylinders. The connecting rods would have been familiar to a Rolls-Royce mechanic, with a similar fork-and-blade design but with roller bearings in the big-end eye of the fork rod and a plain bearing in the blade.

Unlike the Merlin, dry cylinder liners made of steel were screwed and shrunk into the cast silium-gamma alloy cylinder blocks. Dry liners are not in direct contact with the coolant; instead they fit inside cylinders formed in the alloy of the cylinder block. This gives more protection from coolant leaks. Wet liners are a leak waiting to happen, as many early Merlin pilots (and later car owners) found to their cost.

Shrink-fitting is a clever technique using thermal expansion that can be employed by home mechanics. Finding that the steel gudgeon pin is a tight fit in the hole in the alloy piston and won't go in for love nor money, they can place the piston in the oven at high heat and the pin in the freezer (I find the kitchen is always much the best place for engine building). When ready, they will find that the pin slides in easily: the piston has expanded, and the pin has shrunk.

The DB600's huge crankcase was also made of cast silium-gamma alloy. The liners had another clever design feature: they protruded beyond the block and had a screw thread cut around their bottom circumference. This projected into the crankcase and was attached to it inside by large threaded rings that pulled the liners tight against the face of the crankcase, something like a Kilner jam-jar lid. This design saved weight by not requiring dozens of steel hold-down studs and also avoided the possibility of distortion.

The whirling cloud of oil inside the crankcase was prevented from draining down into the cylinders of the inverted engine by the fact that the liners projected above the bottom of the case. Oil was scavenged by suction pumps into the oil tank.

There were other differences from the British engine. As we have seen, the Germans were the leaders in direct injection thanks to Bosch's research into diesel technology, and all the DB601 onwards engines had this feature. The advantages of direct injection into the cylinder are many: they include better distribution to every cylinder, better mixture control and thus power and economy, but most importantly complete immunity from the effects of g, or acceleration in any plane.

As we have seen, this gave the Messerschmitt Bf 109 an advantage over the Supermarine Spitfire, which it was soon to meet during the Battle of Britain. In the early years of the war, the Bf 109 was the main single-engined fighter operated by the Luftwaffe, until the appearance of the Focke-Wulf FW 190. When going into a dive and experiencing negative g, or gravity, the direct injection kept giving the cylinders a positive fuel feed to keep the DB601 running, whereas the Rolls-Royce Merlin with a carburettor would cut out for a few seconds while following the Bf 109. This saved the lives of many Bf 109 pilots, as they were able to escape from pursuing Hurricanes or Spitfires by pushing the stick forwards and power-diving.

The disadvantages of fuel injection were the need for precise machining of the injector pumps and injectors and scrupulous cleanliness of the fuel system, as any diesel-engine mechanic would tell you. German standards were high enough to accommodate these disadvantages. The final disadvantage was the lack of charge-cooling; the Rolls-Royce system of carburation was superior in this respect. Due to the enthalpy of vaporisation or latent heat of evaporation effect, liquid petrol evaporating in the Merlin's inlet tract cooled the incoming air, just as evaporating sweat cools your skin. This enabled more oxygen and fuel to be pumped into the cylinders by the supercharger, thus producing more power.

Another difference between the DB601 and Rolls-Royce Merlin was the supercharger, or blower drive. The German supercharger was hung on the side of the engine – not an ideal place – and was driven by a right-angle drive. The British supercharger was driven from the tail of the crankshaft and fitted neatly at the back of the engine.

The rotational speed of a centrifugal compressor-type supercharger needs to be varied for different altitudes. The problem is that the air-and-fuel mixture is heated by being compressed, just as a bicycle pump will become hot. If we take a supercharger of 2:1 pressure ratio at sea level (that is, it pressurises the inlet manifold at one atmosphere above ambient) the temperature rise of the mixture can be as high as 100° Celsius: the temperature of boiling water. This can reduce the potential power of the engine by around 60 per cent.

The answer was to take off with a reduced amount of throttle and slowly apply full throttle as height was gained and the air grew less dense. At higher altitudes still the supercharger needs to run faster, and later on in the war this was achieved on the Merlin by a two-speed gearbox operated by an altitude-sensing mechanism. What this felt like was described by the Free French Air Force Spitfire pilot Pierre Clostermann, in what has been acknowledged as the finest aviation book to come out of the Second World War: *The Big Show*:

I had just discreetly set course for England when a bunch of Focke-Wulfs decided to take an interest in my poor isolated Spitfire which seemed so ill at ease. Stick right back, 3,000 revs., plus 20 boost, I climbed desperately, followed by the Fw's – two to the right, two to the left, a few hundred yards away. If I could reach the second speed of my supercharger before being shot down I would diddle them.

Six thousand feet. You need about two minutes at full throttle to reach 13,000 feet. In the present circumstances you might just as well have said two centuries.

Twelve thousand five hundred feet – I felt the sweat trickling down the edges of my oxygen mask and my right glove was absolutely sodden.

A roar, and my blower came into action before they could get into firing range. In desperation one of them sent me a burst but without touching me. I now easily drew away and was saved for the time being.[5]

And that is why RAF pilots loved the Rolls-Royce Merlin.

Daimler-Benz designed a supercharger drive that was steplessly variable in ratio. Instead of a choice between two ratios, their supercharger would gradually increase in speed as the air pressure decreased, thereby sustaining a consistent boost in the inlet tract. They did this by building a hydraulic coupling in the driveshaft between the crankshaft and the supercharger impeller. This was fed by two oil pumps, one of which bled off less and less pressure as the aircraft gained height. This was metered by a sleeve valve in the oil flow that was controlled by a barometric cell. The two oil pumps reached maximum output at 13,000 feet (3,962 metres, the same height as Clostermann's Merlin's second stage), and the supercharger's impeller would be turning at

25,000 rpm (the ratio being ten times the rotations of the crankshaft). Frankly this was a better solution. But at that height the DB601 would be producing around 1,175 and the Merlin over 1,300 horsepower.

As the DB601 engine was upside down, the countershaft through which the crankshaft drove the propeller was bored to allow a cannon to be fired through the whole length of the engine. The German design office had not forgotten that the whole purpose of this engine was to carry a gun up into the sky and shoot down an enemy.

Shooting down the enemy was the task of the British pilots during the Battle of Britain, and in particular the hordes of Heinkel He 111 medium bombers that were attempting to wipe out their airfields and cities. The Heinkel was a twin-engined aircraft that, like the Fairey Battle, was obsolete by 1940. As we have seen, it had been designed in 1934, and because of international concern, masqueraded as a civil airliner, being known as the 'wolf in sheep's clothing'. Its two Jumo J211 engines were another liquid-cooled inverted V12 design somewhat similar to the DB600 series, except it was slightly larger at 35 litres and had only three valves per cylinder. With 68,248 manufactured, it was the most numerous German aero engine of the war.

It might seem odd to have employed two such similar engines, but there was limited production capacity for each type and it made sense to maximise factory output. The DB600 engines had a slight edge in lightweight, single-engine applications, which left the Jumo J211 to fill the remaining role as a bomber engine. There is also a danger in relying totally on one design which might be flawed. Perhaps the Germans had heard about the disastrous Dragonfly.

An evolutionary biologist might start to notice similarities between aero-engine evolution and that of biological species.

Might these German liquid-cooled V12s be an example of parallel evolution, with the air-cooled radials taking another evolutionary path altogether? Either way, all varieties of piston aero engines were to be virtually extinct soon, in the blink of an evolutionary eye.

A German air-cooled radial engine that gave the British great cause for concern was the BMW 801 which appeared in the Focke-Wulf 190 fighter in August 1941. This 14-cylinder 41-litre monster developed no less than 1,677 hp. That was a specific power of only 40 hp per litre, but the sheer size of the engine more than made up for this. But the German engine makers had problems. They had poor fuel and a scarcity of vital metals. Fortunately for them they built big light engines that revolved lazily, unlike the highly tuned British engines. These milder engines tolerated the 87-octane fuel that they were sometimes obliged to use. This was called Grade B4; a dark-blue-coloured fuel expensively synthesised from coal. Unfortunately for them this fuel got into the lubricating oil of the Me 109 E and F fighters and thinned it, causing engines to fail. Their high-octane C3 fuel was even worse; reacting with the rubber fuel bags in Me 109 Fs and dissolving into a green slime, once again causing engine failure. Furthermore, the Allied blockade of scarce metals led to a reduction of 50 per cent in the nickel content of German exhaust valves. This led to disastrous engine failures: overheated pistons were burning through in 1½ hours. It took a whole year for the cause to be picked up: BMW weren't talking to Daimler-Benz, and a test engine remained working reliably because no one had thought to fit it with the lower-quality valves! This points to a problem with German management and leads us to an important point: *the British had better bureaucrats.*

The popular image of the British Air Ministry as bumbling civil servants is unfair: wartime Britain was generally better managed than Germany. For example, every factory had a Resident Technical

Officer feeding back information to the Air Ministry. Airfields had Rolls-Royce representatives reporting back from the aircrews. The British kept their key engineers working at what they were good at and didn't sack them for political reasons, whereas the Germans often conscripted them or wasted them on pointless tasks. One of their best engineers, Hans List, was given the job of improving the Mustang's Merlin engine![6] The Air Ministry discouraged engine designs that were too revolutionary, such as the jet or the Sabre, so that work could be concentrated on a simpler, reliable and proven design that could be incrementally made better and better: the Merlin. The British had better planning, resourcing and direction. And in the end the German engines tended to blow up. The Rolls-Royce Merlin didn't.

The British Air Ministry had realised the importance of fuel as early as 1935, when they commissioned a synthetic fuel factory at Billingham in case of interruptions in the supply of crude-oil. And from January 1937 they had been convinced of the need for 100-octane petrol.

The first cargo of 100-octane rich-response fuel was shipped on the SS *Beaconhill* tanker from the Lago refinery on the island of Aruba in the Caribbean, arriving in Britain in June 1939. It was followed by many more deliveries. In early 1940 all Rolls-Royce Merlins installed in Hurricanes or Spitfires were converted to 100-octane. The effect was dramatic: the engines could now tolerate 12 lb of supercharger boost, increasing their power from around 1,060 hp to over 1,300 hp. That 30 per cent increase made all the difference during the Battle of Britain in a closely fought conflict: the Messerschmitt Bf 109E was lighter than either the Hurricane or the Spitfire, and as we have seen it had a bigger engine, but now the Merlin could overpower it. Rolls-Royce had to keep developing the Merlin over 52 marks with better supercharger design, better fuels and better mechanical components to withstand the extra power, just staying ahead of the DB601 engine. This was nothing less than a horsepower race.

France had collapsed in little over a month and sought an armistice with Germany. It was clearly going to be Britain's turn next. At this awful moment in the nation's history Winston Churchill gave his 'Finest Hour' speech in the House of Commons on 18 June 1940. His position in Parliament was shaky, as his appointment was against the wishes of the King and of many Conservative MPs who would have preferred the Foreign Secretary, Lord Halifax. Churchill knew that his oratory was going to be his best weapon.

His speech began with a description of the disaster suffered by the Norwegians and the French, then asked: 'But the great question is, can we break Hitler's air weapon?' He asserted that the RAF, though less numerous than the Luftwaffe, was still powerful and had proven itself 'far superior in quality, both in men and in many types of machine …' He had typed the speech out, revising it up to the last minute, and the last passage was laid out in a blank-verse format reminiscent of the Psalms:

What General Weygand has called the Battle of France is over.

I expect the battle of Britain is about to begin.

Upon this battle depends the survival of Christian civilisation.

Upon it depends our own British life, and the long continuity of our institutions and our Empire.

The whole fury and might of the enemy must very soon be turned on us.

Hitler knows that he will have to break us in this island or lose the war.

If we can stand up to him, all Europe may be freed and the life of the world may move forward into broad, sunlit uplands.

But if we fail, then the whole world, including the United States, including all that we have known and cared for, will sink into the abyss of a new dark age made more sinister, and perhaps more protracted, by the lights of perverted science.

Let us therefore brace ourselves to our duties, and so bear ourselves, that if the British Empire and its Commonwealth last for a thousand years, men will still say, 'This was their finest hour.'

This was one of the great speeches in the history of rhetoric, on a par with Henry V's 'We few, we happy few'. Like Lawrence of Arabia's, Churchill's style was somewhat high-flown, and before the war might have seemed orotund. But at last recent events had become serious enough to be worthy of Churchillian oratory. He made repeated use of groups of three phrases and then contrastive pairs, e.g.: the three battles of France, Britain and the battle for Christian civilisation. Then the pairs: 'Hitler knows that he will have to break us in this island' and 'or lose the war'. Then 'move forward into broad, sunlit uplands' contrasted with 'sink into the abyss of a new dark age'.

The 'broad, sunlit uplands' is a political trope which has been revived in recent times. Churchill was an avid reader of H. G. Wells, whose lecture 'The Discovery of the Future' refers to 'the uplands of the future'. It has echoes in John Buchan, too. Wells also wrote *The War in the Air*, in which he declared that armed struggle was shifting 'from the many to the few'. This, as well as *Henry V*, may have inspired Churchill's immortal reference to RAF pilots as 'the few'.

It is an interesting point: these pilots in many ways resembled medieval horse-mounted knights engaged in single combat.

The effect of Churchill's speeches on the general public was inspirational. A Gallup Poll conducted in July gave Churchill an extraordinary 88 per cent approval rating from the public, and in October, despite the London Blitz, he reached 89 per cent.

More sophisticated members of society were less impressed. George Orwell, writing after the war, had heard it all before:

> Political language – and with variations this is true of all polit-
> ical parties, from Conservatives to Anarchists – is designed to
> make lies sound truthful and murder respectable, and to give an
> appearance of solidity to pure wind.[7]

Churchill repeated his speech on the radio that night while smoking a cigar, which, together with his trouble with the letter 'S', gave the impression that he was drunk. One woman said that he sounded like a bishop.

The Battle of Britain was fought on the production lines as much as in the skies. Rolls-Royce workers received a message from their general manager, Ernest Hives: 'Work till it hurts!' And it did hurt: illness and absenteeism figures rose. But production of Merlins was increasing now, thanks to the Crewe factory coming on stream. More factories were needed and Ford had been approached in the USA to make Merlins under licence. Henry Ford initially agreed to do so, and his son was delighted, as a Rolls-Royce contract to build their engines would be quite an endorsement. But his father then reneged on the deal, stating openly that he thought Britain would be defeated by Germany.

Henry Ford had curious views. He was given to cynical remarks such as 'an idealist is a person who helps other people to be prosperous', and 'I wouldn't give a nickel for all the history in the world. It means nothing to me. History is more or less bunk.' His sympathy for Germany and dislike of Jews was evident: 'International financiers are behind all war. They are what is called the

International Jew: German-Jews, French-Jews, English-Jews, American-Jews … the Jew is the threat.'

He put his money where his mouth was with a publication distributed in Germany: *The International Jew, the World's Foremost Problem*. He also paid for 500,000 American copies of the anti-Semitic fabricated text *The Protocols of the Elders of Zion*, which purported to prove Jewish plans for global domination. Adolf Hitler mentioned Henry Ford favourably twice in *Mein Kampf* and kept a life-sized photograph of him next to his desk, saying: 'I shall do my best to put his theories into practice in Germany', and modelling the production lines of the Volkswagen, the people's car, on those of the Model T. Henry Ford's ignorance of history may have led him to believe that he had invented mass production: he did not, that was the Chinese.

It must have been disappointing for Rolls-Royce when Ford withdrew support for building the Merlin in the USA. However, they had also approached the Packard Motor Company, who accepted an initial order of 9,000 engines. This company had experience of building aero engines, having made the disappointing Liberty L-12 engine and then their own Packard 1A-2500. A full set of engineering drawings was handed over by the Ministry of Air Production, but oddly they first had to be converted from the British first-angle projection convention to the American third-angle projection. When a three-dimensional component such as a crankshaft is ordered from a manufacturer, two-dimensional drawings have to be made.

This is part of the genius of *Homo faber*. A series of orthographic drawings can contain enough information to replicate a chunk of carved metal on the other side of an ocean. They are drawn as if the crankshaft was inside a six-sided cardboard box, with its image projected on the inside walls. The walls are then unfolded to reveal the drawings, and although all six sides could be represented, usually three sides and measurements give enough information to then make the three-dimensional object. These views are named front view, top view and end view. Architects might use elevation, plan and section. In the first-angle projection, the object lies between the observer and the plane of projection, and in the third-angle, the projection plane is transparent and lies in between observer and object.

A sceptic might ask why the Packard factory couldn't just accommodate themselves to the British way of doing things, but in another way they did. The British engine was bolted together using a number of British standard screw-threads, both right- and left-handed, including a lot of British Standard Whitworth threads (BSW). The Packard tooling had to be changed to make this, as the thread standard was not used in the USA. The idea was that the American and British-built Merlins could use interchangeable parts on the production line and also in service on airfields.

Sir Joseph Whitworth was one of Britain's greatest inventors and engineers of the Industrial Revolution. Machines made of metal had to be bolted and screwed together in some way, and a host of local screw standards had arisen, which was hopeless if you wanted to screw a nut made in Knutsford onto a bolt made in Bolton. Whitworth saw the need for a standard thread, and his British Standard Whitworth was the world's first national screw-thread standard.

Whitworth threads enabled the rise of modern mass production, which is the continuous production of standardised and interchangeable parts and is the way the Rolls-Royce Merlin was produced. The Chinese invented mass production around 480 BCE by making interchangeable bronze trigger parts for their crossbows, then the Venetians built ships of standard parts, and of course books were mass-produced from the invention of printing in the fifteenth century.

The British Navy employed mass production in the modern era when in 1803 the engineer Sir Marc Brunel (father of Isambard Brunel) managed to produce 100,000 pulley blocks per year. And during the Crimean War Sir Charles Napier ordered 120 gunboats, each with a steam engine of 60 horsepower, for the campaign of 1855 in the Baltic. They had to be ready in 90 days, and although the gunboat hulls could be built in the time, the steam engines could not. Then the designer of the engines, John Penn, had a brainwave. He took apart a couple of his engines and sent each of the parts to the best machine shops in the country, ordering 90 copies to be made exactly the same as the sample. All the orders were fulfilled in time and he managed to build 90 identical engines in 90 days, partly thanks to the use of standardised Whitworth threads. Along the way John Penn seems to have invented just-in-time manufacturing, too.

Today, 5⁄32 inch Whitworth threads have been the standard toy Meccano thread for years, and stage lighting, camera tripods and sound booms still use BSW, as metric threads were found to be too fine and took too long to screw up when wobbling on top of a ladder. And in the 2011 movie *Cars 2*, bolts fitted to a car with Whitworth threads identify the villain, Sir Miles Axlerod.

Packard rose to the challenge and, like the Ford factory, improved the Rolls-Royce drawings, which turned out not to be quite as accurate as might be expected. This was because the Rolls-Royce Merlin was hand-built: the Derby factory was used to using selective assembly to compensate for the inability of their machine tools to produce tight, repeatable tolerances. The engine builder would have to choose a pair of parts that would fit together with the desired degree of clearance when assembled. And the designers knew that the experienced Rolls-Royce foundrymen would produce castings to the dimensions they *intended* rather than

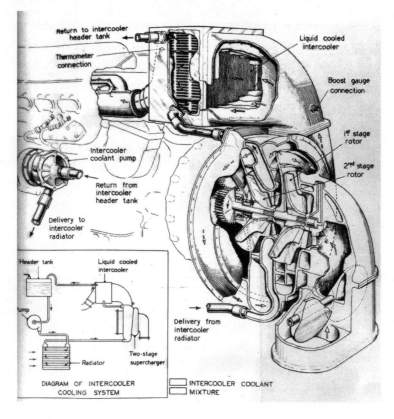

Return to intercooler header tank

Thermometer connection

Liquid cooled intercooler

Boost gauge connection

1st stage rotor

2nd stage rotor

Intercooler coolant pump

Return from intercooler header tank

Delivery to intercooler radiator

Header tank

Liquid cooled intercooler

Pump

Radiator

Two-stage supercharger

Delivery from intercooler radiator

DIAGRAM OF INTERCOOLER COOLING SYSTEM

INTERCOOLER COOLANT
MIXTURE

The two-stage supercharger with intercooling gave the Merlin 60 another 300 horsepower and the Spitfire another 70 mph. The pinnacle of Rolls-Royce supercharger development.

drew. In contrast the Americans with better machine tools were able to achieve interchangeability with huge gains in efficiency. The first Packard-produced Merlins were Merlin 28s, which were the same as the British Merlin XX.*

* Packard also changed the main crankshaft bearings to a silver–lead combination which added indium plating. This improved load-carrying and corrosion-resisting abilities. The Germans, who examined downed Merlins extremely closely, missed this improvement, which they dismissed as an impurity.

The Americans had staggering powers of mass production: Packard produced 2,000 engines per month in 1944 and made a total of 55,523 Merlins. When they were supplied to the U.S. Army Air Forces they were named V-1650, which denoted a V inline engine of 1,650 cubic inches. Lord Beaverbrook and Ernest Hives could at least sleep easier now that Merlin production had been ensured. Now the Germans had to be held off until the Americans could be persuaded into the war.

CHAPTER SIXTEEN

The pity of war

The Battle of Britain was named before it was fought. Winston Churchill's 'battle of Britain' grew a capital letter and entered history as the first major campaign fought entirely in the air.

There have been hundreds of books written about the conflict, all relying on thousands of contradictory reported sightings, shootings-down and kills, all viewed through clouds of propaganda and the shifting fog of war. The Prussian military analyst Carl von Clausewitz had recognised this problem in 1832: 'War is the realm of uncertainty; three-quarters of the factors on which action in war is based are wrapped in a fog of greater or lesser uncertainty. A sensitive and discriminating judgment is called for; a skilled intelligence to scent out the truth.'[1]

The truth we are here to scent out is the part that was played by the Rolls-Royce Merlin engine.

It is clear now that the British knew much more of what was going on thanks to their mastery of radar, the employment of the Dowding System which passed on those radar sightings to pilots, and their breaking of the German Enigma codes. The result was

Battle of Britain artwork of a Hawker Hurricane shooting down a Messerschmitt Bf 109.

the war's first major defeat of Germany's military forces, with British air superiority the key to victory. And the Merlin was the key to the air superiority.

Adolf Hitler had always admired Britain: 'If I had a choice between Italy and England,' he said to a Nazi Party member in 1936, 'I would naturally go with the English … I know the Englishmen from the last war, they are hard fellows.'[2] He often remarked to Albert Speer, his Minister of Armaments and War Production, that the English were 'our brothers. Why fight our brothers?' And the last British ambassador to Berlin, Sir Neville Henderson, wrote that Hitler 'combined … admiration for the British race with envy of their achievements and hatred of their opposition to Germany's excessive aspirations'.[3] In that he resembled Kaiser Wilhelm II.

Hitler had hoped that the British government would negotiate a peace treaty, and when it did not he reluctantly started to make plans for an invasion. On 16 July 1940 he issued Führer Directive No. 16:

> As England, in spite of her hopeless military situation, still shows no signs of willingness to come to terms, I have decided to prepare, and if necessary to carry out, a landing operation against her. The aim of this operation is to eliminate the English Motherland as a base from which the war against Germany can be continued, and, if necessary, to occupy the country completely.

The German code name for the invasion was *Seelöwe*, Sea Lion. The Germans were so sure of an armistice that they began making street decorations for the victory parades of their returning troops. However, Hitler insisted that there must first be air and naval superiority over the English Channel.

German preparations began at once to train troops and to assemble a large number of canal barges and transport ships along the Channel coast. Disturbing plans were made to subjugate Britain after the successful invasion: all the able-bodied male population between 17 and 45 were to be interned and sent to work as slave labour on the Continent. All aero-engine and aircraft research and manufacturing such as the Rolls-Royce factories would be appropriated. SS-Brigadeführer Dr Franz Alfred Six was appointed to direct state police operations in an occupied Great Britain following invasion. His *Einsatzgruppen* death squads were to round up all the Jews in Britain, around a third of a million people, and their extermination would have followed. With Britain unavailable to the USA as an invasion base, the triumph of Germany in Europe would have led to millions of further deaths of people the Nazis disliked.

The squads were provided with a Black Book list of 2,820 people who were to be arrested immediately. It was contained in *The Gestapo Handbook for the Invasion of Britain*, collated by SS General Walter Schellenberg. It contained chillingly accurate information about the British Secret Service, including individual buildings. There were a few strange inaccuracies. It claimed that the Boy Scout movement was 'a disguised instrument of power for British cultural propaganda ... plus an excellent recruiting ground for British Intelligence Services'. There was no mention of the more dangerous paramilitary Brownies. After the war the list was discovered, and it was considered an honour to be in the Black Book. Noël Coward was in it. So was the author Rebecca West, who sent him a telegram saying: 'My dear – the people we should have been seen dead with.' Our *Eminent Victorians* biographer Lytton Strachey was included, but he had died in 1932. And the cartoonist David Low commented when informed he was on the Nazis' death list: 'That's all right. I had them on my list too.'

Franz Six never did become Colonel SSGB. However, he was chief of Vorkommando Moscow, a unit of *Einsatzgruppe* B in the Soviet Union which killed around 200 Jews. Franz Six escaped the death penalty at the Nuremberg trials and was eventually released after serving seven and a half years. He bought a large house near Munich and joined Porsche Diesel as a publicity manager. He retired on a pension and died in his sleep in 1975.

The defending British therefore had a great deal at stake in the prevention of a German invasion of their country, but precious little to defend it with. By the time the Norwegians and the French surrendered the British had lost over 1,000 aircraft: 66 in Norway and 959 in France. Five hundred and nine of these were fighters, which left only 331 Hurricanes and Spitfires to defend the whole of Britain. There were a further 150 outdated fighters, including the Gladiator biplanes. Now only 32 squadrons of fighters were left, instead of the minimum of 52 that the RAF had considered vital for the defence of the country.

Factories all over Britain worked flat out to build more fighters, and by 10 July 1940 there were a total of 527 Hurricanes and 321 Spitfires, with around 600 fighters ready for action. Arrayed against them were three times as many Luftwaffe aircraft.

Imagine, if you will, a Supermarine Spitfire on a grass airfield in July 1940. The other Spitfires and Hawker Hurricanes dotted around the grass share the same Rolls-Royce Merlin III engine driving variable-pitch three-bladed propellers. Two men, the ground crew, are assigned to this particular aircraft. The young pilot, who looks about 20 and comes from a Commonwealth country, comes racing up to the left-hand side and is helped into the cockpit. He is a novice and is muttering to himself the mnemonic known to all Spitfire pilots: 'BTFCPPUR':

brakes, trim, flaps, contacts, pressure, petrol, undercarriage and radiator.

To start the Merlin, the pilot has to turn both fuel-cock levers to On, open the throttle lever half an inch, set the mixture control to rich, pull the airscrew (propeller) control fully back to put it into fine pitch, and make sure the radiator shutter is open.

Next he switches on the magnetos and pumps the Kigass priming pump to spray neat fuel into the inlet manifold. When it starts to pump stiffly he knows the fuel is spraying correctly. This is done because the carburettor cannot provide fuel-and-air mixture while the engine is stationary.

Carrying enough batteries in the aircraft to start the Merlin would limit performance, so the aircrew have a trolley with a set of batteries and a cable that they would plug into the starboard side of the Spitfire.*

'All clear?' the pilot shouts to the two men. They nod, and one puts his thumb up. 'Contact!' A muffled whining from the starter motor, and the airscrew begins to revolve slowly, jerkily. Then, suddenly, with a noise like thunder, the Merlin fires. The exhausts spew long flames enveloped in black smoke, and the whole aircraft starts to tremble with suppressed excitement.

As soon as the Merlin is running evenly the pilot screws down the priming pump. He is ready.

He nods at the ground crew and they drag away the chocks that have been blocking the wheels. The Spitfire taxis carefully away to the runway and the pilot hears his radio announce his call sign: 'You may scramble now; you may scramble now.'

---

* Later marques of Merlin had the extraordinary Coffman 'shotgun' starter, an explosive cartridge of cordite detonating in a cylinder containing a piston mounted on a coarse screw-thread that rotates the engine.

He lowers his seat and slowly opens the throttle. And the Rolls-Royce Merlin thunders up to a thousand horsepower.

Air is roaring into the carburettor, where it is mixed with the 100-octane petrol from the Caribbean and sucked into the eye of the supercharger, where it is whirled into the inlet manifolds. The inlet valves open, jabbed by the camshafts, and the fuel mixture is sucked into the cylinders, where it is compressed in a millisecond by 12 aluminium pistons. The magnetos provide sparks at exactly the right moment, the charge explodes and drives down the pistons, which spin the crankshaft, which spins the propeller. Around 14,000 parts are whirling, jerking, pounding, straining to make this aircraft fly, every one of them made with precision by men and women in Rolls-Royce factories.

Fragments of advice race through the young pilot's brain: 'Don't let the nose drop! – you'll hit the propeller …' The Spitfire starts forward and the airfield begins to sweep past faster and faster on either side. 'Keep her straight! …' The pilot pushes the rudder to check incipient swings.

Suddenly he is aloft. A railway line flashes beneath him. Trees. A line of houses.

He raises the undercarriage and slides the transparent cockpit canopy over his head, throttles back the Merlin and adjusts the airscrew control for cruising. The Spitfire feels wonderfully light on the controls.

For the German pilots it was the Air Battle for England – they referred to the United Kingdom as England. Their dates for the campaign are from July 1940 to June 1941, which includes the Blitz. The British recognise the Battle of Britain as lasting from 10 July until 31 October 1940.

Actually, like most historical events it began gradually. In July the Germans began to attack convoys in the English Channel, then in August the Luftwaffe started attacking British airfields. On 13 August they began an assault they named *Adlertag*: Eagle Day. Unlike their terrifying blitzkrieg attacks on the Continent it did not go well for them. Till now they had crushed feeble opposition, and it is possible that the Luftwaffe was overconfident. They lost 75 aircraft on that day to the British 54.

The Luftwaffe were trying to get their bombers over RAF airfields to destroy aircraft and facilities. Their bombers were the Heinkel He 111, the Junkers Ju 87 (Stuka) and the Junkers Ju 88, all with the liquid-cooled Junkers Jumo engines. They also had the Dornier Do 17 with twin Bramo nine-cylinder air-cooled radial piston engines. Neither engine type was powerful enough. The more numerous Hurricanes were detailed to shoot these bombers down, while the faster Spitfires held off the Messerschmitt Bf 109E and Bf 110C fighters.

As we have seen, the Heinkel He 111, with its distinctive wing shape and all-glazed nose, was a twin-engined aircraft that, like the Fairey Battle, was obsolete by 1940. It had been designed in 1934, and by 1940 its bomb load of 4,400 lb (2,000 kg) was not enough to dent the British defences. With a maximum speed of around 273 mph (440 km/h) it was too slow to escape the Hurricanes, and with only three machine guns it was insufficiently armed. The Heinkel, like the other German bomber types, was vulnerable to the RAF's fighters, but it had one feature that saved the lives of its five young German aircrew: its structural strength. Many aircraft managed to return to their airfields in France with hundreds of 0.303 bullet holes in their fuselage and flying surfaces. The Dornier Do 17s carried an even smaller bomb load of 1,000 kg, and were also slow and poorly armed. On 18 August no fewer than eight Dorniers were shot down and nine damaged in attacks on RAF Kenley, to the south of London.

The Junkers Ju 87 Stuka dive-bomber that had struck so much terror into Spanish civilians with its screaming sirens was easy meat for the RAF. Again like the Fairey Battle, it was slow and underpowered, with a single Junkers Jumo engine, and so many were shot down that it was soon withdrawn from the air battle.

The Junkers Ju 88 was a more formidable proposition. In 1940 it was the most modern German bomber, first taking to the air in December 1936. However, the design was compromised by the late insistence that it should be capable of dive-bombing. This came from the top: Ernst Udet, deputy to the Luftwaffe's commander-in-chief Hermann Göring. The wings were strengthened, dive brakes were added, the fuselage extended and the number of crewmen increased to four. No fewer than 50,000 modifications had to be made, which delayed production and increased the weight from seven to twelve tons. This resulted in a speed loss of around 80 mph (129 km/h). In the end the Ju 88 proved as vulnerable to RAF fighters as other Luftwaffe bombers during the Battle of Britain.

The crash landing of a Junkers Ju 88 led to the Battle of Graveney Marsh. History books claim that the last battle fought on British soil was the Battle of Sedgemoor or the Battle of Culloden, but this farcical Dad's Army encounter was in fact the last ground engagement between British and enemy forces on the mainland. It happened on Friday 27 September 1940, in Kent. Spitfires and Hurricanes attacked what they recognised as a new model of Ju 88, a variant they had been ordered to try to capture. One Jumo engine had already been damaged by anti-aircraft fire during its raid on London, and a Spitfire neatly took out the other engine. The bomber crash-landed on Graveney Marsh and the aircrew crawled out. A detachment of London Irish Rifles, who somehow had managed to get themselves billeted in the nearby Sportsman public house, wandered down to take a look. To their surprise they came under machine-gun fire and they hastily returned to the pub for their rifles. It was a Friday evening, but still they managed to return accurate fire, enough to hit a German airman in the foot and force a surrender.

The commanding officer, Captain John Cantopher, heard one of the captured crew say in German that the plane should 'go up' at any moment. In an act of great initiative and bravery he raced back to the Ju 88, found an explosive charge under a wing and flung it into a dyke, saving the aircraft and its new bombsight for British experts to evaluate. Everyone, including the Germans, then retired to the Sportsman for a drink before the prisoners were picked up. 'The men were in good spirits and came into the pub with the Germans,' said Corporal George Willis, the regiment's piper. 'We gave the Germans pints of beer in exchange for a few souvenirs. I got a set of enamel Luftwaffe wings.'

The German fighters were a far tougher proposition. The Luftwaffe's Messerschmitt Bf 109E had a formidable reputation, as the Nazi propaganda machine had made sure it was in the public eye. It made its debut during the 1936 Berlin Olympics when the proto-type was flown over the stadium. Bizarrely, Messerschmitt first flew this prototype with a Rolls-Royce Kestrel engine. Messerschmitt had exchanged four Rolls-Royce Kestrel VI engines for a Heinkel He 70 Blitz for use as a Rolls-Royce engine testbed, demonstrating the promiscuity of the aero-engine industry in those innocent pre-war days. The admiring crowds were not to know that the engine was made by Rolls-Royce – why should they? But knowledgeable observers might have noticed the high position of the exhaust ports that gave the game away. Then in November 1937 a new world record was set by the Messerschmitt 109 at 379 mph (610 km/h). The implication was that the aircraft was a stand-ard production model, but this was far from true. It was powered by a special version of the DB601 engine, developing 1,660 hp instead of the 1,100 hp of the production version. The engineers did this by increasing the compression ratio from 6.9:1 to 8:1, adding fuel injection and increasing the maximum revolutions to 2,650, but most of all by using special fuel. This was a cocktail of benzol, methanol, ether and alcohol, and completely unsuitable for wartime use due to its expense and instability.

There was another instance of Nazi high-speed propaganda in 1939: the Messerschmitt 109R racing aircraft. This was numbered to cause deliberate confusion with the 109. In fact it was a quite different, tiny aircraft with wings filled with evaporative steam cooling to avoid drag-inducing radiators, and as a result it could never be fitted with guns or indeed survive any damage. At first designated 209V, it captured a new world speed record of 469 mph (754.8 km/h) on 26 April 1939, a piston-engine record that stood until 1969. After the record, in a deliberate disinformation campaign it was given the designation Me 109R. This propaganda was intended to give an aura of invincibility to the Bf 109, an aura not dispelled until the Battle of Britain.

The theme of engine swaps between sides continued. During the war a captured Messerschmitt Bf 109 was fitted with a Merlin engine, all RAF personnel concerned agreeing that the aircraft was now much improved by the change. You might expect this kind of verdict, just as when a captured Spitfire was fitted with a German DB605 engine the same conclusion was reached by the Germans!*

At first the German test pilots disliked the Bf 109, and for several good reasons: the steep ground angle, with the engine sticking up at the sky, resulted in poor forward visibility when taxiing; the sideways-hinging cockpit canopy could not be opened easily to escape a flaming cockpit; the narrow-track undercarriage was

---

* The last version of the 109 was built in Spain after the war and also featured a Rolls-Royce Merlin engine. The Israelis bought Bf 109s and used them in combat in 1948, ignoring the irony of a Nazi-designed aircraft defending the Jewish homeland.

unstable; and the clever automatic leading-edge slats were considered dangerous. Events bore these fears out – in tight turns the spring-loaded slats would flutter, causing instability and the risk of a stall. The Spitfire and the Hurricane could always turn tighter than a Bf 109. But the German fighter was fast and light, and it could fire a heavier cannon through the inverted engine compared with the British fighters' lighter wing-mounted machine guns.

The RAF also had to deal with the twin-engined Messerschmitt Bf 110 heavy fighter. As a prototype the 110 had been fitted with Junkers Jumo 210B engines, which with only 690 hp left the aircraft disappointingly slow. The fitment of DB601 engines in time for the Battle of Britain gave a top speed of 336 mph, enough to keep up with the Hawker Hurricanes. But the Bf 110 had a fatal weakness, which was exacerbated by another poor decision by Luftwaffe high command. It was a large, unwieldy aircraft and was easily spotted during daylight raids. It had been designed to range ahead of bombers, clearing the sky of fighters, but pilots were ordered to accompany the slow bombers in close escort. Their lack of agility and poor acceleration made them an easier target for the British fighters. Thirty Bf 110s were shot down on 15 August 1940, with a further 23 lost over the next two days.

Hermann Göring watched the performance of this aircraft with dismay. His own nephew, Hans-Joachim Göring, was killed flying a Messerschmitt Bf 110 on 11 July 1940, when his aircraft was shot down by Hurricanes of No. 87 Squadron. It crashed into Portland harbour. After 18 August there was a marked reduction in the numbers of Bf 110s appearing in British skies: replacements were not keeping pace with losses. The Luftwaffe had started the Battle of Britain with 237 Bf 110s. Two hundred and twenty-three were lost in the course of it, with many young lives.

* * *

At the beginning of the Battle of Britain the Luftwaffe had around 805 serviceable Messerschmitt Bf 109s and around 1,000 service-able medium bombers in a total of around 2,550 aircraft. The RAF fielded 754 single-seat fighters and 149 two-seat fighters. Horsepower was crucially important to the British, both to climb up to meet the incoming bombers and to evade the protecting fighters. This is where the Rolls-Royce Merlin excelled. By using the newly introduced 100-octane fuel, the supercharger boost could be increased from 6¼ psi to 12 psi for five minutes, giving a maximum combat power of 1,310 hp at 9,000 feet (2,743 metres). The pilot pushed aside an emergency boost override stop at the end of the throttle and used everything the Merlin could give for five minutes. Any longer than this risked overheating of crucial components such as exhaust valves and piston crowns, but plenty of pilots during interceptions just rammed through the boost override and let the engine take its chance, a practice that the Merlin engine took remarkably well. This had to be logged on return, however – if you returned.

Significantly, the German pilots were not allowed to override their boost controls with impunity. Examination of downed German aircraft revealed that clockwork devices were attached to the throttles of some aircraft which only allowed a three- or four-minute period of full boost. It then cut out and prevented re-engagement until a further two or three minutes had passed. This may have saved overstressing the engine, but left the pilot vulnerable to British fighters whose access to full boost was left to the discretion of the pilot. Once this information was passed to the RAF squadrons, they could seek opportunities of forcing the enemy to use full power and wait for the cut-out period. They could then pounce.[4]

Overriding the pilot's discretion suggests a cultural difference between the combatants that one can only speculate upon. Is it possible that the Nazis were more dictatorial? Limiting pilot discretion seemed to have been a poor idea on our contemporary Boeing 737 Max.

The DB601A engine of the Messerschmitt Bf 109E could manage only 1,036 hp at 5,250 feet (1,600 metres), using a similar *Kurzleistung*, or short-term output four-minute limit. The main reason was the poorer-quality fuel, and thus an inability to tolerate high supercharger boost. This was a significant advantage to the British side, and it was maintained by the Rolls-Royce Merlin throughout most of the war, thanks mainly to the feverish development work being done back at Derby.

The Free French RAF pilot Pierre Clostermann encountered a Bf 109 in his Spitfire after the Battle of Britain, and his account gives a flavour of the many duels between these aircraft. He was over his native France when he spotted a Bf 109G on 26 September 1943. His Spitfire's Merlin 66 engine now had 1,720 hp and the Messerschmitt's DB605 around 1,455 hp. The difference was telling:

Three minutes, and the dot had become a cross, about 2,500 feet immediately above me. At that height it was probably one of the new Messerschmitt 109 G's. He waggled his wings … he was going to attack at any moment, thinking I had not seen him. In a trice solitude, poetry, the sun, all vanished. A glance at the temperature and I pushed the prop into fine pitch. All set. Let him try it on!

Another minute crawled by. By dint of staring at my opponent my eyes were watering.

'Here he comes!'

The Hun embarked on a gentle spiral dive, designed to bring him on my tail. He was 600 yards away and not going too fast in order to make certain of me.

I opened the throttle flat out and threw my Spitfire into a very steep climbing turn which enabled me to keep my eyes on him and to gain height. Taken by surprise by my manoeuvre, he opened fire, but too late. Instead of the slight 5° deflection he was expecting, I suddenly presented him with a target at 45°. I levelled out and continued my tight turn. The 109 tried to turn inside me, but at that height his short wings got insufficient grip on the rarefied atmosphere and he stalled and went into a spin. Once again the Spitfire's superior manoeuvrability had got me out of the wood.

For one moment I saw the big black crosses on the 109 standing out on the pale blue under-surface of his wings.

The Messerschmitt came out of his spin. But I was already in position, and he knew it, for he started hurling his machine about in an effort to throw off my aim. His speed availed him nothing, however, for I had profited by his previous false move to accelerate and now I had the advantage of height. At 450 yards range I opened fire in short bursts, just touching the button each time. The pilot of the 109 was an old fox all the same, for he shifted his kite about a lot, constantly varying the deflection angle and line of sight.

He knew that my Spitfire turned better and climbed better, and that his only hope was to out-distance me. Suddenly he pushed the stick forward and went into a vertical dive. I passed onto my back and, taking advantage of his regular trajectory, opened fire again. We went down fast, 470 m.p.h. towards Aumale. As I was in line with his tail the firing correction was relatively simple, but I had to hurry – he was gaining on me.

At the second burst three flashes appeared in his fuselage – the impact visibly shook him. I fired again, this time hitting him on the level with the cockpit and the engine. For a fraction of a second my shell bursts seemed to stop the engine. His propeller suddenly stopped dead, then disappeared in a white cloud of glycol bubbling out of the exhausts. Then a more violent explosion at the wing root and a thin black trail mingled with the steam gushing from the perforated cooling system.

It was the end. A tongue of fire appeared below the fuselage, lengthened, licked the tail, and dispersed in incandescent shreds.

We had plunged into the shadows ... a glance at my watch to fix the time of the fight – twelve minutes past five.

As for the Messerschmitt, he had had it. I climbed up again in spirals, watching him. He was now nothing but a vague outline, fluttering pathetically down, shaken at regular intervals – an explosion, a black trail, a white trail, an explosion, a black trail, a white trail ... Now he was a ball of fire rolling slowly towards the forest of Eu, burning away, soon scattered in a shower of flaming debris, extinguished before they hit the ground.[5]

From 13 August to 6 September 1940 the Luftwaffe attempted to destroy the RAF in southern England with bomber raids on airfields, plus night-time bombing of ports and industrial cities. It is generally accepted that due to poor intelligence reports Göring expected the RAF airfields to be overwhelmed in a few days. Instead, the British fighters just kept coming, and the Luftwaffe pilots began to suffer battle fatigue, which they called *Kanalkrankheit* – 'Channel sickness'. They had to face two long flights over water, and as soon as the red low-fuel warning light came on, they knew they had to turn back. If they were shot down the best they could hope for was to become a prisoner. Many of them were taking Pervitin methamphetamine tablets, known as *Stuka-Tabletten* (Stuka-Tablets). These improved performance and wakefulness but also involved side-effects of depression and addiction. The defenders had the psychological advantage of fighting over home turf, and if shot down they could live to fight another day. One British pilot was shot down no fewer than five times during the Battle of Britain, but was able to crash-land or parachute back to friendly territory each time.

All the resources of Fighter Command were stretched to the utmost, and Göring felt that he had nearly achieved his goal when Adolf Hitler suddenly ordered that the Luftwaffe offensive should be switched to bombing London: the Blitz of 7 September to 2 October. This gave vital breathing space for Fighter Command to recover.

The probable reason for this was the operations of the RAF's Bomber Command. The story of the Battle of Britain has been told many times, but this crucial factor has often been missed out: the activities of British bombers during the conflict. The story of the Battle of Britain in popular imagination overstresses the importance of the defending fighter pilots, and this probably began with the huge success of a threepenny booklet published during the war. *The Battle of Britain* was an Air Ministry propaganda publication that sold in large numbers abroad. It concentrated on the bravery of the RAF fighter pilots but failed to mention the RAF bomber attacks on invasion barges and Berlin. Eighty-one RAF bombers flew across the city on 25 August 1940 and bombed residential areas, causing civilian casualties. Hitler was furious, and when there were further RAF raids on Berlin he ordered a Blitz on London. Churchill understood this, as can be seen in this rarely quoted part of his speech of 20 August:

> Never in the field of human conflict was so much owed by so many to so few. All hearts go out to the fighter pilots, whose brilliant actions we see with our own eyes day after day; but we must never forget that all the time, night after night, month after month, our bomber squadrons travel far into Germany,

find their targets in the darkness by the highest navigational skill, aim their attacks, often under the heaviest fire, often with serious loss, with deliberate careful discrimination, and inflict shattering blows upon the whole of the technical and war-making structure of the Nazi power. On no part of the Royal Air Force does the weight of the war fall more heavily than on the daylight bombers who will play an invaluable part in the case of invasion and whose unflinching zeal it has been necessary in the meanwhile on numerous occasions to restrain.

The Battle of Britain had been won by the defending side, and this conflict can be seen as the turning point of the Second World War. With Britain able to act as a huge aircraft carrier off the shores of Europe the military power of the USA could be directed upon the Third Reich until it was defeated.

Would Hitler's invasion have succeeded? Probably not. A war game conducted at Sandhurst in 1974 concluded that the RAF would have withdrawn its fighters to Midlands airfields, out of range of the Luftwaffe, which therefore could not gain air superiority. And the Royal Navy was powerful enough to destroy the German invasion fleets. All six umpires declared the invasion a resounding failure.

There were better engines than the Merlin, but they didn't win the Battle of Britain, and the Daimler-Benz engines can hardly be blamed for losing it. Just as during the Schneider Trophy, when the best engine won the races, during this conflict the best engine won the war. If any aero engine can be said to have won the Second World War, it is hard to think of any candidate to rival the Rolls-Royce Merlin.

CHAPTER SEVENTEEN

Perfection is finally attained not when there is no longer anything to add, but when there is no longer anything to take away.[1]

The de Havilland Mosquito, together with the Supermarine Spitfire and the Avro Lancaster, was one of the three best British aircraft of the Second World War. It is no coincidence that all three were powered by the Rolls-Royce Merlin engine. What is extraordinary about the Mosquito though – the fact that set it apart from all other combat aircraft it faced – was that it was made of wood and stuck together with glue made of milk.

Lightness is a most desirable characteristic in any dynamic machine such as a car or an aeroplane. Adding lightness entrains a wonderfully beneficial circle: a lighter airframe needs less substantial wings, fuselage, wheels and undercarriage. This in turn makes the whole aircraft lighter, which requires an even lighter airframe. If you then fit a pair of powerful engines such as the Merlin you end up with a high-performing featherweight such as the de Havilland Mosquito.

Geoffrey de Havilland was one of Britain's foremost aviation pioneers and aeronautical engineers. He was responsible for

Leading Aircraftwoman G. Batchelor checking a Mosquito engine.

several First World War aircraft, and between the wars his factory built the Moth series of civil biplanes. He designed and built the wooden DH91 Albatross high-speed airliner and eventually the world's first jet airliner: the Comet.

In the late 1930s he realised that a fast, lightweight unarmed bomber could be highly effective in the approaching war. By using the wooden construction methods perfected in his Albatross, and the twin-engined format of his DH 88 Comet, he knew he could expect faster climb rates and a higher top speed than any pursuing fighters. This would remove the need for heavy guns, thus building in lightness with all its concomitant benefits. Wood also allowed for a perfectly smooth aerodynamic surface free of rivets and seams, particularly when covered with a thin layer of doped fabric.

At first the Air Ministry was dubious, maintaining that wooden aircraft belonged to the past. So de Havilland had to fund the Mosquito as a private venture, just as Vickers had funded the Wellington bomber, Hawker funded the Hurricane, and Rolls-Royce funded the Merlin. But after the Battle of Britain de Havilland's arguments prevailed and 20 bomber variants and 30 fighters were ordered.

In fact the Mosquito represented the future. Whereas the Boeing B-17 Flying Fortress carried three tons of machine guns, ammunition, and turrets, plus the gunners themselves in their heavy heated suits and oxygen cylinders, and suffered the drag of gun barrels, open waist hatches and turrets, the Mosquito relied on being fleet of foot. And so it was the first of a new generation, whereas the B-17 was the last of the old. The Mosquito led directly to the British Canberra and the nuclear V-bombers, none of which mounted a single gun.

The brilliance of the de Havilland Mosquito was that because of its lightness and the power of the Merlins it could also perform photo-reconnaissance, anti-submarine, shipping-attack and night-fighter roles. Mosquitos were used by the RAF Pathfinder Force, which found and marked targets for night-time strategic bombing. The Mosquito became the greatest all-rounder of the war.

Curiously, the balsa-cored construction foreshadowed that of modern sailing yachts. The fuselage was an oval-section mono-coque with no frames, made in two vertically split halves like an Airfix model. First mahogany or concrete moulds were made, then 3-ply birch plywood veneer skins were laid in place, then a layer of glue, followed by an Ecuadorian balsawood core, then another layer of plywood glued and ironed over that. Steel bands then applied pressure. The resulting skin sandwich fuselage was only just over half an inch thick (14 mm).

The layers of wood were glued together with casein, an adhesive made from milk, and fastened together with 50,000 tiny brass screws. The de Havilland company pioneered the use of radio-frequency heating to accelerate curing of the adhesive: a forerunner of the microwave oven. The Mosquito was built by a pool of skilled cabinet makers, car coachbuilders and piano makers, helping to increase wartime production using non-critical workers. Britain was desperately short of imported wood veneers, though. Several men and women could work on each half-fuselage together, speeding up production. The one-piece wing was made of Alaskan spruce, English ash and Canadian birch, with the laminated spars reaching from wingtip to wingtip with no central joint. A cotton-Bakelite composite, 'Tufnol', was used as washers to form load points. The wings were covered in Indian cotton fabric and doped in the same way as the Hawker Hurricane. This was no less than an organic airframe.

The pilot and the observer of the Mosquito sat side by side, as in the front seats of a car, and enjoyed perfect forward vision. Two

Merlin 21 two-speed single-stage supercharged engines were installed, one either side of the cockpit, driving three-bladed de Havilland Hydromatic constant-speed controllable-pitch propellers. The Merlins had a combat power of 1,490 hp at 3,000 rpm, using 16 psi boost, at 12,500 feet (3,810 metres).

The results exceeded de Havilland's wildest hopes. The prototype easily outpaced the accompanying Spitfire. The original calculations suggested that as the Mosquito was twice the weight, had twice the surface area and twice the power of the Spitfire, then it should only be a matter of 20 mph faster. In fact the prototype was quite a bit faster than the Spitfire Mk II, which had a top speed of 360 mph (579 km/h) at 19,500 feet (5,944 metres). That Mosquito managed a top speed of 392 mph (631 km/h) at 22,000 feet (6,706 metres), making it one of the fastest aircraft in the world. When it was fitted with Merlin 61s it was faster still. These Merlins now had two-speed, two-stage superchargers and the two-piece cylinder block and heads which had at last caught up with the accelerating production. Power was now 1,560 hp at 3,000 rpm, with 15 psi boost at 11,250 feet (3,429 metres). The first flight with the new engines was on 20 June 1942, and the prototype hit 428 mph (689 km/h) at 28,500 feet (8,700 metres), about the height of Mount Everest.* Even if a Merlin failed the Mossie could return home on one engine at over 200 mph.

Sergeant Mike Carrick remembered the new aeroplane being delivered to his base near Norwich one day in 1942:

---

* An RAF Mosquito belonging to RAF 684 Squadron, based at Alipore airfield, Calcutta, made an 'accidental' flight over Mount Everest in 1945. The 400 feet (120 metres) of 35 mm film shot of the mountain helped route-finding on the successful British expedition of 1953. The altitude ceiling of this Mosquito was 37,000 feet (11,278 metres).

On 15 November it came suddenly out of nowhere, inches above the hangars with a cracking thunderclap of twin Merlins. As we watched, bewitched, it was flung about the sky in a beyond belief display for a bomber that could outperform any fighter. Well-bred whisper of a touch down, a door opened and down the ladder came suede shoes, yellow socks and the rest of Geoffrey de Havilland.* [It was an] impossible dream of an aircraft … It had awesome power on the leash in those huge engines and was eager on its undercarriage like a sprinter on the starting blocks who couldn't wait to leap up and away.†

For around two years there was no enemy aircraft that could catch the Mosquito, and the photo-reconnaissance versions roamed the length and breadth of Axis-held Europe with impunity. They quickly supplanted the Spitfires for this role because they could carry more cameras, they had a longer range, they had the security of two engines, the navigator could locate and identify the targets, and the later versions could fly at extremely high altitudes.

A single Mosquito flew a reconnaissance sortie over Berlin in March 1943 and was ineffectually pursued by several Me 109s and FW 190s which failed to catch it. This may have influenced Hitler's order to switch development of the revolutionary Me 262 jet aircraft towards a *Schnellbomber* (fast bomber). But the technology was too late for Hitler: 'Look on my Works, ye Mighty, and despair!'

* * *

* This was de Havilland's son, also named Geoffrey. Sadly, two of de Havilland's sons died testing his aircraft: John, who died in an air collision involving two Mosquitos in 1943, and Geoffrey, who was killed in 1946 flying the jet-powered DH 108 Swallow while diving at the speed of sound.

† This aircraft, W4064, was to be shot down six months later on the squadron's first operation.

The Rolls-Royce Merlins 23s in some night-fighter versions of the Mosquito were modified to use 'laughing gas' nitrous oxide injection. The decomposition of the N_2O gas inside the cylinders provided extra oxygen for burning more fuel, adding 360 hp for a maximum of six minutes, the same as an extra Eagle engine. Two squadrons of Mosquito XIIIs were fitted with the system, which provided 30 lb of the gas per minute. This increased the speed of the Mosquito by 47 mph at 28,000 feet (8,534 metres).

These night-fighter aircraft were painted black, and it was found in April 1942 at the testing facility at Boscombe Down that a smooth black finish instead of matt black increased top speed by 10 mph.[2] Just carefully cleaning the aircraft added 3 mph to the top speed.

What was the Mosquito like to fly? Whereas the twin-engined Messerschmitt Bf 110 heavy fighter had been a cumbersome disappointment, the twin-Merlin-engined de Havilland Mosquito was a lightweight delight. However, the high power-to-weight ratio and high wing loadings made the Mosquito a nervous thoroughbred and a handful to fly. On take-off the sheer power of the long, out-thrust engines produced a great deal of yaw, or twisting on a vertical axis.

You are sitting in the left-hand seat ready for take-off with your navigator sitting close next to you. You are a highly experienced pilot, but both of you know that the full fuel load and the 4,000 lb Blockbuster bomb makes the Mosquito tail-heavy and tricky to get into the air. This is a dangerous load: the shockwave makes the safe dropping height 6,000 feet (1,829 metres).

You get the go-ahead and release the brakes. Leading with the port engine and carefully opening the throttles helps, but the Mosquito at once starts to veer to the left. It doesn't have a locking tailwheel to hold a heading during the first part of the take-off roll, so you have to use differential braking to catch the take-off swing.

The Merlins are roaring at full power now, and you silently pray they keep running for the rest of the flight. Once rolling you have learned to ignore the official take-off speeds and accelerate hard along the runway no matter how long it is, pulling up only right at the end.

Once in the air there is a long dangerous period of gathering airspeed during which a Merlin failure would be fatal, the aircraft rolling over and plunging into the ground. The V_{MC} (minimum control speed) – the speed that needs to be reached to assure rudder effectiveness with one engine feathered and the other running at full power – is a sobering 172 mph, the highest of any Second World War twin-engined aeroplane. Below that the good engine would have to be quickly throttled back, but the power would then be insufficient to keep your heavily laden Mosquito in the air.

Once safely in the cruise you notice once again that the Mosquito needs only the lightest of control forces, which remain unusually light at high speeds. Most other aircraft are self-limiting; their controls become heavy at speed and make it difficult for a clumsy pilot to pull the wings or tail off. Your Mossie rewards sensitive handling.

Only the most talented and experienced pilots were selected for the Mosquito, and as a result the aircraft gained a great deal of kudos and affection. At one squadron the aircrew used to go out and share an evening drink with their Mossie.

As a result of the abilities of aircraft and crew the Mosquito gained a reputation for precision low-level bombing attacks, and became famous for daring daylight hit-and-run raids such as the attacks on the Gestapo headquarters in Oslo, Aarhus and Copenhagen. On 31 January 1943, Mosquitos attacked the Grossdeutscher Rundfunk state radio station in Berlin when Hermann Göring was due to address the nation to mark the tenth anniversary of the Nazi accession to power. The raid kept him off the air for an hour, and the sounds of confusion were broadcast all over the Continent. Six weeks later, berating an audience of German aircraft manufacturers, Hermann Göring said:

> It makes me furious when I see the Mosquito. I turn green and yellow with envy. The British, who can afford aluminium better than we can, knock together a beautiful wooden aircraft that every piano factory over there is building, and they give it a speed which they have now increased yet again. What do you make of that? There is nothing the British do not have. They have the geniuses and we have the nincompoops. After the war is over I'm going to buy a British radio set – then at least I'll own something that has always worked.[3]

The most famous and controversial Mosquito raid of all was probably Operation Jericho, of 18 February 1944. This was a daylight low-level precision-bombing raid on Amiens prison which was planned to break down the walls, burst open the cell blocks and enable the escape of prisoners. The attack had been requested because many of the French *résistants* were due to be executed the next day by the Gestapo. They had been successfully locating the hidden V-1 flying-bomb launch sites, and the British wanted to demonstrate support. The attack was supposed to have been led by Air Vice-Marshal Basil Embry, but an intercession by Air Marshal Trafford Leigh-Mallory* resulted in a last-minute change – Embry knew too much about the forthcoming invasion on the Normandy beaches. And so at the last minute the raid was led by Group Captain Percy Charles Pickard, who, although hugely experienced

---

* The brother of George Mallory of Everest, he was killed in an air crash in the Alps in the November of that year.

in night-time bombing had limited experience of daytime low-level raids with a Luftwaffe presence.

Pickard had become something of a national celebrity after the release of *Target for Tonight*, a 1941 British documentary film about the crew of a bomber taking part in a raid over Germany. As a propaganda film it had enormous popular success, but most of the flight officers and crew who appeared in the film failed to survive the duration of the war. Pickard was one of them.

Eighteen Mosquito Mk VIs with 1,635 hp Merlin 25s took off into the worst winter weather any of the pilots had known. Flying low over snow-covered France, following the straight Albert–Amiens road, the Mosquitos inched lower still. Now, ten feet off the ground, the propellers whirled twin snowstorms off the fields behind each aircraft:

> … and there, a mile ahead, was the prison. It looked just like the briefing model and we were almost on top of it within a few seconds. We hugged the ground as low as we could, and at the lowest possible speed; we pitched our bombs towards the base of the wall, fairly scraped over it – and our part of the job was over.[4]

The Mosquitos had to be under the height of the prison walls when they released their bombs, then they had to climb fast enough to clear them. This was when the throttle-responsive Merlins earned their keep. Without the lag of turbochargers or jet turbines the supercharged Rolls-Royce engines could react instantaneously to demand.

The bombs blew holes in the walls, destroyed guard houses and broke down both ends of the prison. The governor, Eugene Schwarzenholzer, was decapitated. Pickard was at the end of the second wave, checking that the damage was sufficient. He was jumped by a pair of Focke-Wulf 190s, one of which shot his tail off. His Mosquito crashed seven miles north-east of the prison.

Back in Amiens 102 of the 717 prisoners were killed, 74 wounded and 258 escaped, including 79 Resistance and political prisoners, although two-thirds of the escapees were recaptured. Two Mosquitos and two escorting Typhoons were shot down.

Was it all worth it?

Controversy about Operation Jericho arose after the war. Who exactly ordered the raid? Why was a letter sent from 'C', the head of the Secret Intelligence Service, thanking the RAF for its execution of the raid? Was it something to do with two recently captured Allied prisoners who knew about D-Day? Further investigation suggests that any wartime suppression of the news had more to do with the loss of the popular hero Pickard than with any kind of conspiracy.

These daring daylight raids made the Mosquitos visible to the citizens of occupied countries in a way that the night-flying heavy bombers were not, and this gave them hope that the Nazis were not indestructible. Wing Commander John Wooldridge gave the flavour of a Mosquito attack:

It would be impossible to forget … the sensation of looking back over enemy territory and seeing your formation behind you, wing-tip to wing-tip, their racing shadows moving only a few feet below them across the earth's surface; or that feeling of sudden exhilaration when the target was definitely located and the whole pack were following you on to it with their bomb doors open, while people below scattered in every direction and the long streams of flak came swinging up; or the sudden jerk of consternation of the German soldiers lounging on the coast, their moment of indecision, and then their mad scramble for the guns; or the memory of racing across The Hague at midday on a bright spring morning, while the Dutchmen below hurled their hats in the air and beat each other on the back. All these are unforgettable memories. Many of them will be recalled also by the peoples of Europe long after peace has been declared, for to them the Mosquito came to be ambassador during their darkest hours.[5]

Advertisement for the de Havilland Mosquito, *circa* 1943. Already looking ahead for post-war sales.

CHAPTER EIGHTEEN

*For they have sown the wind, and they
shall reap the whirlwind.*[1]

Perhaps the most unexpected achievement of the Rolls-Royce Merlin engine was to transform a disaster of an aircraft into a triumph: the Avro Lancaster bomber. The Germans themselves considered this to be the best night bomber of the war, and quite apart from thousands of sorties over Germany the Lancasters carried out the famous Dambuster raids and succeeded in finally sinking the *Tirpitz* battleship.

Churchill was convinced of the need for heavy bombers. In a letter to Beaverbrook, the Minister of Aircraft Production, he wrote on 8 July 1940:

> When I look round to see how we can win the war, I see that there is only one sure path. We have no continental ally which can defeat the German military power ... Should [Hitler] be repulsed here or not try invasion, he will recoil eastward, and we have nothing to stop him. But there is one thing that will bring him back and bring him down, and that is an absolutely

An Avro Lancaster silhouetted against flares, smoke and explosions during an attack on Hamburg in 1943.

devastating, exterminating attack by very heavy bombers from
this country upon the Nazi homeland.

Three months later he wrote in a memo to Cabinet: 'The Navy can
lose us the war, but only the Air Force can win it. The fighters are
our salvation, but the bombers alone provide the means of victory.'

The story of how the Avro Manchester became the Avro Lancaster
curiously parallels that of the development of the Merlin engine
itself, and indeed the performance of the Allies during the course
of the war: a poor start, with an initial series of disasters overcome
by dogged perseverance, leading to final success. And if the Rolls-
Royce Merlin was a triumph of development over design, the
tragic story of the Avro Manchester shows what happens when an
aero engine is insufficiently developed.

In contrast to the Mosquito, the Manchester was an example of
an aeroplane that just couldn't help putting on weight. It was
designed to fulfil the British Air Ministry's Specification P.13/36,
which called for a twin-engine monoplane medium bomber capable of carrying out dive-bombing attacks, lifting heavy bomb loads
or two 17-foot (5.26-metre) torpedoes. Also, it had to be capable
of catapult-assisted take-offs.* The dive-bombing, torpedoing and
catapult demands all added structural strength and weight that
couldn't be removed when these demands were eventually
dropped, rather like the Ju 88. Furthermore, the pair of Rolls-
Royce Vulture engines arrived over a ton heavier than promised,
so the anticipated 26,000 lb operating weight swelled to 45,000 lb.

The unfortunately named Vulture was a disaster of an engine. It
was conceived in 1935 when Rolls-Royce realised that they would

---

* It was thought that bombed airfields would be effectively shorter, requiring
catapult-assisted take-off.

need a large new motor for the big twin-engined bombers then being planned. In keeping with their philosophy of building on existing technology, they used the Peregrine V12 design, in effect turning one V12 upside down and putting a second V12 on top of it, resulting in an X24 configuration. The 24 pistons shared a common crankshaft, which had six throws on seven main bearings like the Merlin. However, unlike the Merlin, four pistons and cylinders had to share each big-end bearing, and therein lay the problem.

Whereas Henry Royce, in designing an Eagle X16, had used two big-end bearings per crankpin; two pairs of fork-and-blade rods like those of the Merlin, and staggered the cylinders, the chief designer Albert Elliott opted for a 'star-rod' design, with a master and three articulated rods. Herein lay the defect of the design. The

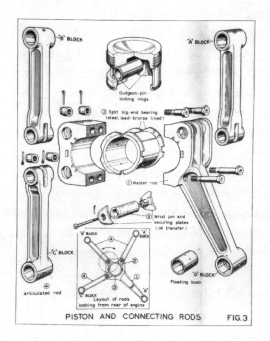

PISTON AND CONNECTING RODS   FIG.3

The heart of the problem: the Vulture big end.

star-rod system not only meant there were different stroke lengths, but also the rods were arranged in such a way that in order to close the big-end bearing onto the crankpin two long and two short bolts had to be used. The loading on the bolts was high and a brittle steel was used, and this proved to be an Achilles' heel.

In operation the Vultures proved barely powerful enough to get the Avro Manchester off the ground, and when they did one engine would often throw out a rod that would virtually cut the engine in half and start a fire. As a particularly cruel refinement, the Vulture would also cut the hydraulic line to the propeller, meaning that it could not be feathered (turned edge-on to the wind). The pilot would suddenly find himself with a dead engine, an engine fire, plus a huge airbrake on the affected wing in the shape of an unfeathered propeller.

Twin-engined aircraft are notoriously hard to fly with one bad engine, and pilots used to say that the only purpose of the good one was to take you to the scene of the accident. The Avro Manchester could not maintain height on one Vulture and the aircraft would slowly and ineluctably lose height. The inevitable result was a crash onto the ground, usually fatal.

There were other design failings in the Vulture. Oil frothing was a constant problem that ought to have been picked up on the test bench and that caused further big–end-bearing problems. Another mistake was the omission of a balance pipe between the two coolant pumps, which caused airlocks, overheating and further engine failures.[2] At first the Vulture was rated at 1,760 hp, but it proved so unreliable that the maximum revolutions were reduced and it was derated to 1,480 hp. Around 30 Manchesters were lost due to engine failure, and many young aircrew died.

Rolls-Royce was desperately busy developing the Merlin in time for the Battle of Britain, so it is perhaps unsurprising that the company devoted less time to an engine intended for a later stage of the war. Roy Chadwick, Avro's chief designer, realised the problems with the Vulture engines early on, and suggested a solution: to fit four Rolls-Royce Merlin XX engines on longer wings. On the positive side, it was acknowledged that the Manchester had good stability and handling, it was easy to fly and it had well-balanced controls. The Air Ministry agreed, and the Avro Lancaster was born.

By now a 27-litre Merlin was developing much the same power as a 42-litre Vulture, and the Avro Lancaster was a revelation, with twice the power of the Manchester and far greater reliability. It was also easy to fly and fast for its size. One former Manchester pilot said that it was like stepping into a sports car after a saloon. Experienced Lancaster pilots were often able to outmanoeuvre Luftwaffe fighters, and there were instances of Lancasters having been looped and barrel-rolled, both intentionally and otherwise. The first ATA woman pilot to fly a Lancaster, Lettice Curtis, remembered that:

they were so easy to fly that pilots were not even required to do a 'stooge' preliminary trip. They had little tendency to swing on take-off, had a simple and well-designed cockpit layout, were no heavier on the controls than, say, a Wellington and most of all were very easy to land because, unlike the Halifax or Stirling, they flew and landed without a significant change of attitude.[3]

Roy Chadwick had provided a 10-metre (33-foot) long, unencumbered bomb bay for the initial torpedo requirement of the Manchester, and the Lancaster with much the same fuselage was able to carry an enormous bomb load, indeed twice the weight of its more glamorous American cousin, the Boeing B-17 Flying Fortress.

* * *

617 Squadron made the new Lancaster famous. It was formed to attack the three major dams that contributed water and power to the Ruhr industrial region in Germany: the Möhne, Eder and Sorpe. Special bombs, named 'Upkeep', had been designed by the engineer Barnes Wallis to bounce over the protecting torpedo nets. These were actually depth charges, as they were triggered by water pressure, using three hydrostatic pistols. They were spun backhandedly at 500 rpm by a Vickers hydraulic turret motor and dropped at 60 feet and 230 mph onto the surface of the water, whereupon they bounced across the intervening nets and hit the dam's inner wall, rolling down and detonating. Sir Arthur 'Bomber' Harris, the wartime head of Bomber Command, was not convinced, remarking that the bouncing bomb was 'tripe beyond the wildest description. There are so many ifs and buts that there is not the smallest chance of it working.'

A secret type of Lancaster was built, the 'Type 464 Provisioning'. A suffix (G) meant these aircraft had to have an armed guard at all times. The bomb-bay doors were removed and struts to carry the huge cylindrical bomb were fitted. These would swing apart at the moment of release. Lamps were fitted in the bomb bay and nose which when shone on the water converged at a height of 60 feet. The mid-upper turret was removed to save weight and drag, and the gunner moved to the front turret so that he could assist with map-reading.

On the night of 16/17 May 1943 nineteen of these special Lancasters took off, and after a low-level flight in moonlight across Germany they attacked and breached the Möhne and Eder dams. Around 1,600 civilians and prisoners of war were drowned in the tumultuous floods. Eight Lancasters were lost and 53 of the 133 aircrew killed, a casualty rate of almost 40 per cent. Harris was unconvinced: 'The destruction of the Mohne and Eder dams was to achieve wonders. It achieved nothing compared with the effort and the loss.' However, Albert Speer, the German Minister for Armaments, said that the raid was 'a disaster for us for a number of months' and Germany had to divert 20,000 labourers from building defences in France into repairing the dams, something that had a significant impact on D-Day. The audacious and technically challenging raid also had huge propaganda value and helped British morale at a difficult time. It has now entered British popular legend, particularly after the 1955 film *The Dam Busters*. It is worth mentioning that the Merlins ran faultlessly throughout.

617 Squadron were to have another famous success with the Lancaster: the sinking of the *Tirpitz*. The German battleship was a thorn in the side of the Allies as she lay in a Norwegian fjord acting as a 'fleet in being' – a threat that extended a controlling influence without leaving port. Large Royal Navy units were tied up in the area just in case *Tirpitz* broke out. If she had met SS *Queen Elizabeth* or *Queen Mary* troopships in mid-Atlantic carrying 15,000 American servicemen, she could have sunk them in 30 minutes and changed the course of the war.

Tirpitz was a monster, the biggest battleship ever built in Europe, with nearly 50,000 tons in displacement and with 13-inch armoured sides. She would be difficult to deal with.

The Royal Navy had sunk her sister ship *Bismarck* in 1941, but *Tirpitz* had survived one attack by British mini-submarines and a series of large-scale air raids by the Fleet Air Arm. So the RAF was given the job of sinking her.

Barnes Wallis had designed another earth-shattering bomb: the 6.5-metre (21-foot), 4.5-tonne green-painted Tallboy, designed to attack hardened targets such as submarine pens. Unlike standard thin-skinned bombs, these had to penetrate a long way before exploding, so the casing was cast in one piece of hardened steel. A man then had to crawl inside and hammer the interior fair with a pneumatic tool – a deafening task. Tallboys were expensive and slow to produce, having spent weeks cooling after their Torpex explosive had been melted in kettles and poured into the casing. It was realised that Tallboy bombs could be the answer to *Tirpitz*'s foot-thick steel plating.

Once again Avro Lancasters had to be modified to carry a Barnes Wallis superbomb. The B1 Special Lancasters had their upper dorsal gun turret removed, together with the front guns, the flare chute and the armour plating behind the pilot's head. An extra 250-gallon fuel tank was squeezed in. Special Rolls-Royce Merlins were fitted, too, T-24s, which had 1,610 bhp of take-off power instead of the standard Merlin XX's 1,390. By then Rolls-Royce (following Bristol's lead) had developed a unitised Merlin XX-series engine installation and nacelle, so that the 'power egg' could be speedily swapped after disconnecting pipes, four bolts and the Cannon electrical connecters. Together with new paddle-bladed propellers, these Merlins were now powerful enough to lift Tallboy into the air.

Three attacks on *Tirpitz* were mounted by Nos 9 and 617 Squadrons. On the first attack they were deployed to a staging point in Russia, as the *Tirpitz* was too far north to reach and return safely on one flight. One of the Lancasters staggered back into the air so slowly that it hit a tree and collected a Russian branch in the bomb bay.* On 15 September 1944 they attacked *Tirpitz* and one of the Tallboys penetrated the forecastle and hull, exploding beneath the starboard bow. This wrecked the bows and the forward compartments were flooded with 2,000 tons of water. From then on *Tirpitz* was doomed.

---

* This branch can still be seen in Petwood Hotel, Woodhall Spa, 617 Squadron's original Officers' Mess. *The Lancaster and the Tirpitz* by pilot Tony Iveson describes this whole operation in detail.

The battleship was patched up and moved to Tromsø on 15 October, but the RAF attacked again. These raids were more straightforward than the first, as the port was now within the range of Lancasters flying from airfields in northern Scotland. The second raid caused only minor damage to the battleship, but during the third and last raid *Tirpitz* was struck by several Tallboy bombs, rolled over and capsized with heavy loss of life among her crew.

Barnes Wallis realised that the modified Merlins allowed even heavier loads to be carried, and so he came up with the 9-tonne, 26-foot (8-metre) Grand Slam earthquake bomb. Even more weight was stripped out of the bombers; this time two of the rear guns were removed together with the bomb-bay doors. In the air the load was so heavy that the Lancaster's wings bent upwards 8 inches, and when the Grand Slam was dropped the aircraft shot up 300 feet (91 metres). This monster of a bomb became the most powerful non-atomic aerial bomb used in combat. Falling at nearly the speed of sound it could burrow through 20 feet (6 metres) of concrete and made a crater 100 feet (30 metres) across. A Grand Slam was so valuable that if the Lancaster could not find the target it had to bring it back, landing – rather carefully – at the longer and wider Carnaby runway.

The Avro Lancaster had one fatal flaw: it lacked a belly gun turret. The Luftwaffe developed an upwards-pointing cannon for their night fighters that they called *Schräge Musik* (literally 'slanted music'). The fighter would creep close under the Lancaster from behind, outside the crew's field of view and field of fire. It would then open fire with the cannons, often detonating the bomb load, which took the bomber and sometimes the night fighter with it.*

---

* The epithet has a strange echo of the First World War German machine-gunners' gruesome name for the dance of troops being hit by Spandau MG08 machine-gun fire: '*Spandau ballet*'. The band took their name from Berlin graffiti.

It's a fine June evening on a Bomber Command airfield in East Anglia in 1943. You are sitting on your parachute in the pilot's seat in an Avro Lancaster bomber, and in the briefing you have just learned of your destination: the Ruhr valley, the centre of German industry. A young woman tractor driver has already backed her train of dark-green bombs under your aircraft and they have been winched up through the open bomb doors into the yawning black belly. Your machine, C for Charlie, is one of the oldest on station and one of the first Lancasters built, but her Rolls-Royce Merlins are bang-on. The Squadron's average oil consumption is 13.2 pints per hour, but your motors are all below that because you have a wizard ground crew. Your petrol tanks have been filled with juice, enough to get to the Ruhr and back tonight at a gallon per mile, plus 10 per cent.

You're the first of your flight to start up. You slide open the window and call out to your ground crew: 'Clear for starting?' Twenty feet below, the starter battery trolley is plugged in and ready. You point at the port-side inner Merlin and give the man the thumbs up. He points his left hand at the same engine and revolves his right index finger as if winding up a toy plane. You press the electric starter button and hear a booster pump chattering, then the slow puffing of a Merlin turning over. There are a couple of bangs as individual cylinders ignite, then a roar as all twelve catch fire. One by one you start the other three engines and the racket of Merlins fills all four corners of the fenland evening.

Out on the runway your wireless operator has flashed C for Charlie's code sign, and now you get a green light from the control tower. Your right hand pushes open all four throttles, keeping the port outer slightly ahead to counteract the swing, and C for Charlie starts to roll. The flight engineer is ready to take over the throttles, as you will need both hands to pull back on the control column.

The old Lanc is loaded well beyond peacetime limits, and it's a struggle to get her into the air. 'Ninety-five knots.' By now your Lancaster is thundering towards the end of the runway. Over 5,000 Merlin horsepower are trying to wrestle 60,000 lb of aircraft, bombs and petrol off the ground. 'One hundred knots.' The end of the runway is coming towards you. You touch the rudder bar with your foot, as the port throttles aren't quite far enough ahead. Your navigator has been calling out your airspeed and when he reaches 105 you are suddenly airborne. You lift the dark nose to the sky.

Anyone used to today's comfortably pressurised, high-altitude jet airliners would not believe the discomfort and danger of the Avro Lancaster bomber. As you peer into the darkness over Germany the whole aircraft is roaring, rattling and shaking with the vibration of four high-revving Merlin engines. The instruments are shuddering, and the needles and figures are a blur. With the Merlins running unsynchronised at slightly different revolutions, the racket is unbearable, so you sight through the dazzling blurs made by the airscrews and superimpose them as you juggle the throttles. Watching the stroboscopic effect enables you to synchronise the engines. You have heard stories that Luftwaffe bomber pilots are instructed to desynchronise their engines on approach to the target to maximise the terrorising effect on civilians. You have to breathe oxygen through a mask, and shout instructions and warnings to the rest of the crew using a microphone and headphones. The Lancaster is at the limit of her altitude, and at this height the air is full of turbulence and pockets, so she bucks, rolls and yaws all around the night sky. There are no powered controls and so you have to use all your strength to keep her straight and level. You know that after a dive to escape the ever-present German night fighters the flight engineer would have to help you to pull back the column.

And the first thing you know about the attack is the explosion of incendiary shells in your cockpit as a Ju 88 night fighter rakes the old Lancaster's belly from nose to tail.

Lie in the dark and listen,
It's clear tonight so they're flying high
Hundreds of them, thousands perhaps,
Riding the icy, moonlight sky.
Men, materials, bombs and maps
Altimeters and guns and charts
Coffee, sandwiches, fleece-lined boots
Bones and muscles and minds and hearts
English saplings with English roots
Deep in the earth they've left below
Lie in the dark and let them go
Lie in the dark and listen.
Lie in the dark and listen.[4]

Fifty-seven thousand two hundred and five aircrew in RAF Bomber Command were killed out of a total of 125,000, a 46 per cent death rate. Five thousand were killed in training. Their average age was 21. Between 350,000 and 500,000 German civilians were killed by Allied bombing. Each of the 7,377 Avro Lancasters and their Merlins cost an average of £58,974. Complete with bombs and crew training the whole enterprise cost around £120,000 per aircraft, or around £5m in today's money, and half of them were lost.

There were some astonishing escapes, though. One such was the survival of the Lancaster tail gunner Flight Sergeant Nicholas Alkemade. On the night of 24 March 1944 his Avro Lancaster was attacked by a German Ju 88, and soon the fuel tanks were burning fiercely. The gun turrets were such a tight fit that there was no room for Alkemade's parachute, and when he tried to retrieve it from the fuselage he was met by a gout of flame. Preferring to die by falling than by burning, he returned to his turret, rotated it and jumped out into the night sky. Later he reported that his fall from 18,000 feet (5,500 metres) was peaceful, although he lost consciousness.

He woke up sitting in a snowbank beneath a stand of pine trees. His fall had been broken by hitting branches and then a good depth of soft snow. He was able to move his limbs and seemed to be suffering from only a sprain. He lit a cigarette and waited to be captured. The Germans refused to believe his story until they examined the metal clips to which his parachute would have been attached. They were still sewn in place.

Sir Arthur Harris, head of RAF Bomber Command, was prone to Old Testament language. He described the Avro Lancaster as his 'shining sword'. At the beginning of his bombing campaign he had written:

> The German people entered this war under the rather childish delusion they were going to bomb everyone else, and nobody was going to bomb them. At Rotterdam, London, Warsaw and half a hundred other places, they put their rather naive theory into operation. They sowed the wind, and now they are going to reap the whirlwind.

His campaign to obliterate Hamburg was named Operation Gomorrah. In some post-war assessments Harris has been described as a war criminal and in some as a hero. He was known by the press as Bomber Harris, and privately within the RAF as Butcher Harris.

* * *

Rolls-Royce Merlins were fitted to more and more aircraft as power and reliability increased and production ramped up, and the Handley Page Halifax was another four-engined bomber which started out as a twin-engined design powered by the unhappy Vulture. The Halifax was involved in infiltrating secret agents or Joes into occupied Europe by parachute through their Joe holes, or hatches. The problem was exfiltrating them or picking them back up again. Agents had to return by dangerous land-escape routes, or be flown out by light aircraft, such as the Westland Lysander, that landed on moonlit nights in remote fields.

According to the CIA,[5] the British developed a secret way of scooping an agent off the ground using a Halifax flying at low level. After the war an American inventor, Robert Edison Fulton Jr, tried to copy this technique. He attached a weather balloon to a pig, as pigs have a nervous system and physiology that resembles humans. The low-flying pick-up aircraft caught the balloon and whipped the poor animal off the ground. The pig began to spin on the cable as it flew through the air and it was winched into the aircraft uninjured but somewhat aggrieved. Once inside it promptly attacked the aircrew.

The American P-51 Mustang was another averagely performing aircraft that was transformed by the Rolls-Royce Merlin engine. As a result it became one of the best fighters of the war, possibly *the* best. It was commissioned by the British, who in April 1940 were looking for more supplies of fighter aircraft, to be built in America. It was intended to escort British and American bombers over Germany and therefore had to have a long range to reach Berlin and the ability to fight when it got there. The British initially asked the manufacturers, North American Aviation, to produce a Curtiss P-40 Tomahawk design, which only approximately met the British specifications. The company's president, James Howard 'Dutch'

Kindelberger, said he could do better than that, and offered a new design using the American-built Allison V12 engine of around 1,500 hp. The contract was signed and the prototype was ready in only 102 days. It featured a clever new wing section employing laminar-flow airfoils which proved to be extremely slippery.

At first the Mustang I (P-51A) was a disappointment. The underwhelming American Allison V12 engine had only a single-stage supercharger that caused power to drop off rapidly above 15,000 feet (4,572 metres), and this made it unsuitable for combat at the altitudes at which USAAF bombers flew. The first Allisons were turbocharged, a kind of supercharger driven by a turbine placed in the flow of exhaust gases instead of directly from the crankshaft. British engineers had to help their American allies with the advanced metals needed in such turbines. Samuel Heron proved that Vitallium could be cast by removing his Vitallium dentures and showing them to sceptical American engineers who were discussing turbocharger materials.

Then Rolls-Royce's Ronald Harker tested the Mustang:

The first thing I noticed during the flight was that the indicated speed was some thirty miles per hour more at similar power settings than a Spitfire! The ailerons were light and gave a rapid rate of roll; this was one of the areas where the Focke-Wulf had an advantage over the Spitfire. The guns, too, were close inboard, which gave a concentrated fire, but most important, the internal fuel capacity was three times greater than the Spitfire* – which meant that the Mustang would be able to provide an escort to the bombers for deep penetration into Germany. I

---

* The Mustang had large internal fuel tanks which, with external drop tanks made of paper (!), would enable it to escort USAAF bombers into Germany and back. The fuel capacity was actually 150 imp gallons, whereas the Spitfire's internal capacity was 85 imp gallons.

was very impressed with the aeroplane, which seemed to me to be a natural for the two-speed, two-stage Merlin 66, which was just coming into service on the Spitfire IX. I felt that the Merlin 66 would greatly improve the speed of the Mustang, as its full throttle height would be very much greater than with the Allison engine and the thirty or so miles per hour higher indicated speed maintained at altitude would mean an increase in true speed of perhaps fifty miles per hour![6]

In a refreshingly decisive manner, Hives at Rolls-Royce seized on Harker's idea, so did a supportive Air Ministry, and a Merlin was duly tried in the Mustang. This gave a maximum of 1,720 hp compared with the Allison's 1,200 hp, increasing the top speed from 390 to 440 mph (628 to 708 km/h) and the service ceiling to 42,000 feet (12,800 metres). It was found that the Mustang's airframe produced less drag than the Spitfire's: at sea level the Mustang required 480 hp to enable it to fly at 250 mph, whereas the Spitfire V, which was 25 per cent lighter in weight, required an additional 100 bhp.

Using the same engine, the Spitfire IX's top speed at 11,000 feet (3,353 metres) was 384 mph (618 km/h) and the Mustang's 404 mph (650 km/h). But because it was lighter the Spitfire could accelerate faster and could turn more sharply: pilots were still losing consciousness on tight turns.

The Mustang was closely associated with the American 332 Fighter Group, a corps of black pilots who had trained at Tuskegee, Alabama. The US military was segregated by race during the war; and white and Afro-American personnel were kept separated. The Tuskegee airmen's top scorer was Lee Archer Jr, with five air kills, but one kill was later reallocated to another pilot to prevent Archer being recognised as an ace. He had managed to shoot down three Messerschmitt Bf 109s in ten minutes.

The Mustang was highly manoeuvrable. In the spring of 1944 the 23-year-old US pilot Bill Overstreet was flying his 'Berlin Express', a Mustang escorting bombers. He chased off a pursuing Messerschmitt 109 and damaged its engine. The German pilot flew low over Paris, hoping to get support from the German flak units based there. Unable to shake off the Mustang and growing desperate, the German pilot flew through the arches of the Eiffel Tower, not expecting the Mustang to follow. Bill claimed that he did so, shooting down the Messerschmitt and escaping down the course of the River Seine. Later he was awarded the *Ordre national de la Légion d'honneur* by the French Ambassador to the United States. However, fellow pilots disputed his story, pointing out that he did not mention his exploit until after the war, he gave no date, there was no encounter report, no victory claim, and there was no Messerschmitt 109 crash site.

They didn't dispute the fact that the Mustang was capable of flying under the Eiffel Tower, because there was a photograph of a Mosquito doing exactly that on 14 September 1944, flown by Warrant Officer Bob Boorman (RCAF) and his navigator, Flight Sergeant Bill Bryant (RAF).[7] The Mosquito's wingspan was 17 feet (5.2 metres) greater than the Mustang's. Unlike Overstreet the British aircraft's crew tried to keep it quiet.

When Hermann Göring saw Mustangs over Berlin he was reported to have said that at last the game was up.[8]

The Rolls-Royce Merlin was also made in a de-tuned land version called the Meteor. When fitted into a Crusader tank the test team had trouble recording the top speed with their equipment as it was so fast. This proved the concept, and thousands of tanks were fitted with the Meteor. Early engines were rebuilt from Merlin engines recovered from crashed aircraft.

CHAPTER NINETEEN

Less is more

The Rolls-Royce Merlin might have won the Battle of Britain, but it wasn't necessarily the best aero engine of the war. By late 1941 the Spitfire V was facing the formidable Focke-Wulf 190 fighter, which mounted the air-cooled BMW 801C engine of 42 litres. This was descended from American Pratt & Whitney engines that BMW had built under licence. It had 14 cylinders in two rows and developed 1,539 hp. It had only two valves per cylinder, but it was light. The FW 190 quickly proved superior to the Spitfire in everything except turn radius: the British fighter could still pull tighter blackout-inducing turns.

The FW 190 had been designed by Kurt Tank, who in conversation with Ronald Harker after the war revealed how the aircraft nearly never happened. He had been caught by two Spitfires while flying an Me 108 from France back to Germany. They shot off an aileron, forcing him to land in an open field. He then said: 'Why did they not press home their attack and finish me off? You should find out who the pilots were; they should be reprimanded!'[1]

Production of the 'Corncob', the radial Pratt & Whitney R-4360 Wasp Major. The piston aero engine had become a mechanical monstrosity.

The Rolls-Royce engineers responded to the FW 190 with an even more powerful Merlin engine. Having proved that the basic engine was strong and reliable, the Derby men now concentrated on ramming more and more air-and-fuel mixture into their engine. Rolls-Royce had come up with another idea in the form of the two-stage supercharger. In this design there are two impellers working in series (one after the other). Air would enter the first impeller, be compressed and then be led to the second impeller rotating on the same shaft, which would compress it still further. As we have seen, the original Merlin had a single-stage supercharger and could not hope to compress the air/fuel mixture more than four times. The two-stage blower nearly doubled this and increased the power developed in the thin air of 30,000 feet (9,144 metres) by 42 per cent. What this meant is that at this height the Spitfire IX was no less than 78 mph (125 km/h) faster than the Spitfire V.

As we know, with the high pressure being achieved by the two-speed two-stage blowers the temperature rise of the air/fuel mixture was even more of a problem. Rolls-Royce now came up with an aftercooler, a heat exchanger that removed the excess heat in the compressed mixture and got rid of it in a separate radiator. Thanks to this and the temperature drop due to petrol evaporating in the carburettor, the Merlin was eventually making 1,420 hp at 30,000 feet (9,144 metres) and the Spitfire was able to outpace the Focke-Wulf 190.

In this evolutionary arms race the Germans responded with their HZ-Anlagen system. In their twin-engined high-altitude reconnaissance/bomber Dornier 217 PV1 they put an extra, third, engine inside a fuselage just to drive the supercharger. The service ceiling was an astounding 53,100 feet (16,200 metres), well above any fighter aircraft. And in 1941 Roy Chadwick proposed his Avro 684 Stratosphere Bomber, which was essentially a Lancaster with a pressurised cabin and an extra fuselage-mounted Merlin 45 engine driving a huge supercharger to provide pressurised air to the usual four Merlin XXs. This five-Merlin-engined Lancaster needed no armament and would cruise at 410 mph (660 km/h) and reach 50,300 feet (15,300 metres).* But as with all these developments, the mechanical complexity of piston engines was growing.

Rolls-Royce knew that the Merlin was going to be outclassed soon, and they had a replacement engine waiting: the Griffon. This had been requested in 1938 by the Fleet Air Arm, as they needed a big reliable engine with high power at low altitudes. It was once again a liquid-cooled V12, more than a third larger in capacity than the Merlin at 37 litres. Oddly, it ran in the reverse direction to the Merlin.† The thinking behind this was that any aircraft with a powerplant of around 2,000 hp should be able to have any available comparable engine substituted should the original type fail. Consequently the direction of rotation of the Griffon had to change to match the engines of Bristol, Napier and Armstrong Siddeley (the Merlin, which was being built in the US – whose aero engines all turned the same way as the Merlin – stayed as it was).

The Griffon had to fit in the same airframes as its smaller older brother, so some rather clever work was done on camshaft and magneto drives to make it no longer and not much wider than the Merlin: the twin magnetos of the Merlin were replaced by one driven from the reduction gears at the front of the engine (inside

* Just six years later Chadwick was to design the astonishing Avro Vulcan, a delta-winged nuclear bomber that looked like a spacecraft and made the Lancaster look like a hayrick. All this in just 44 years since the Wrights.

† Two-stroke engines can run forwards or backwards, as moped-owning humorists can find out by bump-starting their scooters backwards. But the reason the first Volvo ÖV 4 off the production line drove backwards with three reverse gears and one forward was that someone put the differential in upside down. This delayed the introduction celebrations by one day. Incidentally, Volvo means 'I roll' in Latin, a reference to their ball-bearing production.

it had two electrically separate circuits for redundancy). And the supercharger was driven from the reduction gears by a long flexible shaft running under the crankshaft. This cushioned snatch-loads caused by rough use of the throttle by inexperienced pilots. As a result of all this work the frontal area of the Griffon was only 7.9 square feet (0.73 square metres) compared with the 7.5 square feet (0.7 square metres) of the Merlin.

Design went much more smoothly than for the Merlin, but the Griffon was too heavy, and although it was ready for testing in June 1940 it did not go into production until March 1942, probably because the Merlin was being developed so fast that it always seemed one jump ahead of its heftier younger brother. By a similar process the Griffon grew a two-stage supercharger, Rolls-Royce fuel injection and produced an immense 2,050 hp. Installed in the final version of the Spitfire it gave a top speed of 454 mph (730 km/h). This last Spitfire was twice as heavy, more than twice as powerful, and climbed 80 per cent faster than the prototype, K5054. Was it nicer to fly? Probably not: the test pilots usually reckoned the Merlin-engined Mark V hit the sweet spot.

The Rolls-Royce Griffon had a long life. It was fitted in the Avro Shackleton maritime reconnaissance anti-submarine aircraft which performed Cold War duties until May 1991, serving the RAF for 49 years.*

So if the Merlin wasn't even the most powerful Rolls-Royce engine, how did it compare with the other Allied manufacturers?

Back in 1920, after the Dragonfly debacle, the Bristol Aeroplane Company had been ordered by the government to take over

---

* Eventually the final version of the Griffon in the 1991 Shackleton developed 2,500 hp on take-off with a life of 2,000 hours. This was the same power as its ancestor the racing R engine, which had a life of one hour.

production of the Cosmos Jupiter air-cooled radial engine. They were reluctant but in no position to argue, and so was born the long line of Bristol radial engines. They perfected the sleeve valve and by the 1940s their giant Hercules was making more power than the Merlin. Having a huge frontal area, it was more suitable for large bombers such as the Avro Lancaster II and Bristol's own Beaufighter. But its power didn't begin to compare with the Napier Sabre.

As we have seen, there had always been bad blood between Rolls-Royce and Napier, going back at least to 1931, when the failing Bentley company had been whipped away from under Napier's nose in an underhand manoeuvre. Despite building the best British engine of the early Thirties, the Napier Lion, Napier had always struggled to make headway in the competitive aero-engine market. Despite all this, they managed to make what was probably the cleverest piston engine of the Second World War.

The engineer Harry Ricardo had decided back in the Twenties that sleeve valves would eventually make more power in aero engines than poppet valves, and went off to work with Bristol. Frank Halford worked in the next office to Ricardo and went off to work with Napier. One afternoon he sat down with H. C. Tryon, Chief Engineer of the tiny Napier company in Acton, and together they designed the Sabre on the back of an envelope. And they designed a masterpiece.

In essence the Sabre was two flat horizontally opposed 12-cylinder engines sitting on top of each other, each with its own crankshaft. These were cunningly coupled together to avoid transient overloads. The 24 cylinders were in two layers; cast in aluminium and bolted to the sides of the crankcase. Inside each cylinder was a sleeve valve, and the rows of upper and lower sleeves were driven from worm gears cut in a shaft running between the layers of cylinders. These shafts were hollow, and down the centre of each ran a torsion bar taking power to the superchargers at the back of the engine – a neat touch.

Here there were no testy exhaust poppet valves to glow red-hot, no nasty constricting inlet valves; the ports in the sleeve valves could allow in more fuel/air mixture than poppets and the combustion chambers could be made into a perfect shape. The constantly moving sleeve meant that cylinder-bore wear was minimised, as it was always moving in relation to the piston, even at the top and bottom of its stroke. And the shallow cylinder heads meant that the engine could be more compact than conventional piston engines with overhead camshafts. This led to a smaller frontal area, which was helpful to the airframe designer.

The Sabre had a capacity of 37 litres compared with the Merlin's 27 litres, with twice the number of pistons. These were therefore smaller, and as a result the Sabre could rev up and down the range like a racing-car engine. The maximum revolutions were 4,200 rpm and the power rating in service was set at 3,750 hp, although it could deliver a reliable 5,500 hp with 45 lb/in of supercharger boost. The most power the Merlin could muster was 2,200 hp at 2,900 rpm, with 2,640 hp available for short periods.

These were incredible figures, but few people today have ever heard of this engine. Much of its history has been obscured by time and (Setright suggests) by the deliberate suppression of records. To his credit Beaverbrook saw the potential of the Sabre when it passed the Air Ministry's 100-hour test and it went into mass production.

It was not an easy engine to make, nor easy to maintain. There was a struggle to make the sleeves perfectly round, but by a piece of luck the Bristol Taurus engine shared the same cylinder bore of 5.0 in. Testing with Taurus sleeves was a success, but initially Bristol refused to manufacture Sabre sleeves, objecting that their techniques were confidential! The Air Ministry had to have a stern word, and in the end the sleeves were manufactured from nitrided austenitic steel forgings using Bristol tooling. Quality was not as good as at Rolls-Royce factories, and engines were often delivered with broken piston rings, and machining swarf left inside. The Merlin was more reliable and far better made. It was also cheaper: costs per horsepower were Merlin: £1, Hercules: £2, Sabre: £5.

By the time the Sabre came into service many of the good pilots had been killed, replaced by recruits of poorer quality and hastier training. They may have been swayed by the effective Rolls-Royce propaganda, and when pilots and ground crew encountered problems with the Sabre they were prone to condemn what they could not understand. And the Sabre's reputation was not enhanced by the Rolls-Royce representatives at the airfields. It didn't have black shiny cam covers like the Merlin, it just looked like a big green metal box with rows of exhaust pipes and sparking plugs.

The Sabre was hard to start from cold, as the sleeve valves were a tight fit. At freezing point the Coffman starter struggled to turn it over, and Pierre Clostermann describes how the ground crews had to start the Sabre in his Hawker Typhoon every two hours during freezing nights in France. The correct technique was to dilute the oil with petrol, a practice also employed by the Germans, but a misfire when starting from cold could result in a fire, and many trainee pilots were terrified of their Typhoons.

The second problem was not the fault of the Sabre. Fighters patrolling the Channel suddenly started having engine failures, and on one day no fewer than 15 Sabres clanked to a halt while up in the air. Frantic investigations revealed the problem, and Setright hints at sabotage:

A bunch of stupid, ignorant and possibly misguided (but by whom?) mechanics had discovered that it was possible to fiddle with the internals of the automatic boost control that integrated the various engine functions … these fighters were cruising up and down the Channel for long drawn out patrols with the airscrews in full coarse pitch, the crankshafts turning over at cruising rev./min., and the blowers were delivering maximum boost! Of course, the engines were hammering themselves to death, and there was practically a riot when the cause was discovered.[2]

The Sabre-powered Hawker Tempest became the most successful destroyer of the German V-1 flying bombs, as it was the fastest of all the Allied fighters. Whereas the Spitfire had to dive to gain enough speed to catch the V-1, the Tempest could actually overtake the primitive cruise missile. Later the Hawker Tempest managed to shoot down 20 Messerschmitt Me 262 jet aircraft.

Imitation is the sincerest form of flattery, and Rolls-Royce produced a copy of the Sabre named the Eagle Mk XXII, with the same layout of 24 cylinders in an H-block configuration and the same kind of sleeve valves. It was never fitted in a production fighter, as something else was coming.

The Napier Sabre was still the most powerful piston engine at the end of the war, and probably the most misunderstood. It was no Italian beauty on the outside, but on the inside it was a paragon of human ingenuity, the apotheosis of the piston engine.

However, the Napier designers were not content to rest on their laurels, and in a virtuoso display of technical brilliance they then determined to make the most fuel-efficient piston aero engine yet: the Nomad. This was to be in competition with the gas-guzzling jets about to displace the piston engine in the civil market. They made a 12-cylindered diesel-fuelled two-stroke, compounded with an exhaust-gas-driven turbine. This turbine was connected to the crankshaft and helped to drive the propeller. Every ounce of energy in every drop of diesel was extracted:

after the exhaust gases were expelled from the cylinders they released more energy inside the turbine. After they left the turbine an afterburner nozzle injected extra fuel into the exhaust flow, using the last wisps of unburned oxygen. The result was the best specific power figures seen up to that time: 0.83 lb/hp compared with the Wrights' 15 lb/hp. But it was not enough to stop the jet revolution.

The piston aero engine eventually became a mechanical monstrosity, and perhaps the most monstrous was the Corncob; the American Pratt & Whitney R-4360 Wasp Major. This was a 28-cylinder, four-row radial piston aircraft engine of no less than 71 litres and 3,500 horsepower: more than twice the size of a Merlin. Each row of seven air-cooled cylinders was slightly offset from the previous row, forming a semi-helical configuration in the vain hope of allowing cooling airflow to the cylinder rows behind them. This shape inspired the engine's nickname. In service the engine suffered tremendous overheating problems that necessitated frequent changes of cylinder, and it proved ruinously expensive to operate. Corncobs also regularly caught fire. Six were fitted to the Convair B-36 Peacemaker intercontinental bomber, plus four jet engines, giving rise to the B-36's slogan 'six turnin' and four burnin''. After the regular Corncob fires this was changed to 'two turning, two burning, two smoking, two choking, and two more unaccounted for'.

In January 1930, two years before Henry Royce signed off the PV12 prototype, a young cadet at RAF Cranwell applied for a patent for a new kind of aero engine that would eventually make the Rolls-Royce Merlin obsolete. In fact, Frank Whittle's jet engine would make nearly all piston aero engines obsolete, and if it hadn't been for a lack of imagination at the Air Ministry and obstructions put in his way by a senior engineer the British could have had jet aircraft before the war.

In 1928 the five-foot-tall 20-year-old RAF Cranwell college cadet submitted a thesis. In it he described a motor-jet; a conventional piston engine driving a compressor to a combustion chamber whose exhaust was used for thrust, something like the Napier Nomad. The idea was not new, but after further reflection Whittle realised that the piston engine was superfluous: a turbine in the exhaust flow could drive the air compressor. He literally threw away the piston engine. Later he expressed his reasoning:

> Reciprocating engines are exhausted. They have hundreds of parts jerking to and fro, and they cannot be made more powerful without becoming too complicated. The engine of the future must produce 2,000 hp with one moving part: a spinning turbine and compressor.

What followed though was a tragic story of official incomprehension, lumbering bureaucracy, the jealousy of seniors and simple snobbery. Whittle was an ex-apprentice in a class full of ex-public-school boys, and he was short and bumptious. He showed his engine idea around his next base, RAF Hendon, where it attracted the attention of Flying Officer Pat Johnson, who had worked as a patent examiner. Johnson took Whittle's idea to the commanding officer of the base. This started a chain of events that almost led to the British jet engine being produced much sooner than it did. Unfortunately, a respected engineer, Alan Griffith, had published a paper which suggested that gas turbines were impractical and better used for turboprops, where the output is used to drive a propeller. The RAF rejected Whittle's idea.

'Whittle peddled his invention from door to door for five years,' Johnson said later. 'Unable to find anyone who was interested, he failed to pay renewal on the original patent taken out in 1930 and was about ready to give up.'

Whittle didn't have the £5 necessary to renew his patent in 1935. But his original patent had been spotted by the German jet pioneer Hans von Ohain, in a German library, although he denied it to Whittle, who believed him. He said later: 'We felt that it looked like a patent of an idea … We thought that it was not seriously being worked on.' He too suffered official intransigence, but persevered, and so the Germans were the first to produce an operational flying jet aircraft.

Just before the outbreak of war Frank Whittle built his engine and demonstrated it to the Air Ministry in a 20-minute, high-power run. At last they woke up and placed an order. On 15 May 1941 it first flew in an airframe specially built for it, the Gloster E28/39. The venue was Whittle's alma mater, RAF Cranwell. But it was too late. The Messerschmitt Me 262 was operational by 1944. Its top speed was 560 mph (900 km/h), whereas the fastest Merlin-powered opponent, the Mustang, could manage only 440 mph (708 km/h). This was a step change in performance.

The advantages of the jet engine were clear to see to an ordinary bystander. In appearance it was a light alloy tube with one rotating part inside: the turbine. Instead of vibrating and roaring like a piston engine it ran smoothly with a gushing sound. Comparisons of power are hard to make, as jet engines provide thrust directly instead of going through an inefficient propeller, but a Boeing 747 needs around 120,000 horsepower to get airborne, the equivalent of about 116 Merlin I engines.

Turbines have high power-to-weight ratios because they can spin at very high speeds. The Space Shuttle had pumps driven by turbines that were about the size and weight of a car engine at 775 lb (352 kg), but produced 72,000 horsepower (53.6 MW), a ratio of 0.01 lb/hp, compared with the Wrights' 15 lb/hp. And jet turbines are more reliable than piston engines because there is less to go wrong.

In a conversation with Whittle after the war, Von Ohain, the German jet pioneer said:

If you had been given the money you would have been six years ahead of us. If Hitler or Göring had heard that there is a man in England who flies 500 mph (800 km/h) in a small experimental plane and that it is coming into development, it is likely that World War II would not have happened.

EPILOGUE

The Rolls-Royce Merlin was designed and built by the finest engineering company in the world. It could serve as the golden spike that marks the moment that the human brain was overtaken by the speed of technology: the Spitfire's tight turns caused her pilots to black out, and those turns were driven by the sheer power of the Rolls-Royce engine. The Wright brothers' 1903 Flyer had an engine of only 12 horsepower: barely that of a lawnmower. Within 30 years the first Rolls-Royce Merlin Mk I was developing 1,000 horsepower, and by 1945 the Merlin RM17SM had been flight-tested at a continuous 2,200 hp, with 2,640 hp available for short periods. This is a rate of progress comparable to our computer industry today. Surely soon computers will displace fighter pilots, and later they will probably displace most of humanity. The Merlin might serve as some kind of marker. This remarkable engine turned the tide of the war during the Battle of Britain, and then Merlin-engined bombers destroyed the Nazi war machine.

After the war, versions of the Merlin were used in civil aircraft such as the Avro Tudor, a development of the Lancaster, and the Canadair North Star, a development of the DC-4. The noise and vibration were not tolerated by paying passengers, though, and the new breed of aero engine took over. The supercharger expertise at Rolls-Royce, learned the hard way on the Merlin, put the company in pole position to make jet engines, and it still does.

The world's first jet airliner, the British-built de Havilland Comet, made its first scheduled flight on 2 May 1952, flying to Johannesburg via Rome, Beirut, Khartoum, Entebbe and Livingstone. The flight, taking just over 23 hours, was a pleasure. Passengers reported a smooth journey, flying high above bad weather en route. There was superb service and beautifully presented meals. The noise and turbulence of the old piston-engine planes was history; this was the new way to travel, the new travellers were the Jet Set, and this was the beginning of the Jet Age.

IN MEMORIAM

Ring out the thousand wars of old,
Ring in the thousand years of peace.[1]

Although the Rolls-Royce Merlin represented a high-water mark of technological achievement it has to be remembered that it was a weapon of war, and directly caused hundreds of thousands of deaths of young pilots, men, women and children. The incendiary bombs that turned Dresden into an inferno were lofted into Germany's night sky by Merlins. We can only hope that the European unity that followed the war can be followed by a lasting peace. That would be the best epitaph for a beautiful piece of machinery.

ACKNOWLEDGEMENTS

As a boy I was shown how petrol engines were put together by kind adults: Jeffery Stockall, industrial chemist; Malcolm Withers, journalist; and Paul Brewer, radar engineer. The writings of L. J. K. Setright furthered my education, and this book could never have been written without access to the library of my good friend Bryan McGee, Rolls-Royce owner and Club member. I must acknowledge that authors receive huge help from our local libraries, and Bristol Central Library has been wonderful in that respect. Thanks to my partner Gina Waggott for her support and encouragement, and to my editor Hazel Eriksson for her calm guidance. Thanks, too, to the patience and good humour of Myles Archibald, who is the publishing director of William Collins, and to my literary agent, the tireless Charlie Viney.

NOTES

INTRODUCTION

1. Setright, *The Power to Fly*. The best of technical writing.

CHAPTER ONE

1. Joseph Needham, *Science and Civilisation in China*, Volume IV, Part 2, quoting the *Baopuzi*, 320 CE.
2. The *Zizhi Tongjian* (*Comprehensive Mirror to Aid in Government*), 1084 CE.
3. R. A. B. Mynors, R. M. Thomson and M. Winterbottom, *William of Malmesbury, Gesta Regum Anglorum*, Oxford Medieval Texts (1998–9).
4. Dee, *The Man Who Discovered Flight*.
5. Rudyard Kipling, *Something of Myself: For My Friends Known and Unknown*, London, Macmillan, 1951 (first published 1937). And Langley was correct in his forecast.
6. Setright, *The Power to Fly*.
7. Howard, *Wilbur and Orville*.

CHAPTER TWO

1. Timothy Leary.
2. C. H. Gibbs-Smith, *Sir George Cayley's Aeronautics 1796–1855*, Her Majesty's Stationery Office, London, 1962. Referred by https://www.aerosociety.com/media/4862/sir-george-cayley-the-invention-of-the-aeroplane-near-scarborough-at-the-time-of-trafalgar.pdf
3. Quoted in Leonard S. Hobbs, *The Wright Brothers' Engines and Their Design*, Washington, DC, Smithsonian Institution Press, 1971.
4. Harvey H. Lippincott, *Propulsion System of the Wright Brothers*, in Howard S. Wolko and John D. Anderson (eds), *The Wright Flyer, an Engineering Perspective*, Washington, DC, Smithsonian Institution Press, 1987.
5. Howard S. Wolko and John D. Anderson (eds), *The Wright Flyer, an Engineering Perspective*, Washington, DC, Smithsonian Institution Press, 1987.
6. From Tom D. Crouch, *The Bishop's Boys: A Life of Wilbur and Orville Wright*, New York, W. W. Norton & Company, 2003.

CHAPTER THREE

1. https://airandspace.si.edu/exhibitions/wright-brothers/online/classroomActivities/8-12_excerpt.cfm
2. *New York Herald*, 10 February 1906.
3. *L'Aérophile*, 11 August 1908.
4. *L'Auto*, 9 August 1908.
5. *St. Louis Post Dispatch* on 7 November 1943.

CHAPTER FIVE

1. C. P. Snow, *The Two Cultures and the Scientific Revolution*, Cambridge University Press, 1959.
2. Quoted in Hodges, *Alan Turing*.
3. Quoted in Strachey, *Eminent Victorians*.
4. Ed Glinert https://ilovemanchester.com/why-mr-rolls-and-mr-royce-probably-didnt-meet-at-the-midland-hotel/
5. Nockolds, *Magic of a Name*.
6. Pugh, *Magic of a Name*.
7. Quoted in Pugh, *Magic of a Name*.
8. Frank Lord, Henry Royce obituary, *Autocar*, May 1933.
9. Lloyd, *Rolls-Royce: The Growth of a Firm*, Macmillan, 1978.
10. According to Rolls's biographer, Lord Montagu of Beaulieu.
11. Lord Brabazon of Tara, *The Brabazon Story*. Private publication, 1956.
12. Lord Montagu, of Beaulieu, *Rolls of Rolls-Royce*, Cassell, 1966
13. Ibid.
14. Kenneth Grahame, *The Wind in the Willows* 1908.
15. Lord Montague, *Rolls of Rolls-Royce*.
16. *The New York Times*, 23 August 1908.
17. Tom C. Clarke, *Royce and the Vibration*, *Damper*, 2003, ISBN 1-872922-18-X.
18. *Country Life* 1905, quoted in Pugh, *Magic of a Name*.

CHAPTER SIX

1. What he actually wrote was 'Le secret des grandes fortunes sans cause apparente est un crime oublié, parce qu'il a été proprement fait: The secret of great fortunes without apparent origin is a crime forgotten, for it was properly done.' Honoré de Balzac, *Revue de Paris, Volume 12, Le Père Goriot*, 1834.
2. Erwin Panofsky, 'The Ideological Antecedents of the Rolls-Royce Radiator', Proceedings of the American Philosophical Society, 1963.
3. Charles Freeston, *Autocar* magazine, 1913.

CHAPTER SEVEN

1. Rolls-Royce Board Minutes, 7 August 1914.
2. Cousin of Sir John, later Lord Montagu. Rolls-Royce always managed to be associated with aristocracy.
3. Bentley, *My Life and My Cars*.
4. Jackson, *Infamous Aircraft*.
5. Quoted in Taulbut, *Eagle*.
6. Rolls drove the 80 hp Mors car at the Duke of Portland's Clipstone Park at 82.8 mph and claimed the kilometre world record in 27 seconds.
7. Jeremy Archer, *From Dorset Yeoman to Distinguished Airman – The Story of Wing Commander Louis Strange*, Dorchester, The Keep Military Museum, 2016.
8. Hoyland, *Last Hours on Everest*, reveals what probably happened to George Mallory on Everest.
9. Lewis, *Sagittarius Rising*.
10. Ricardo, *The Internal Combustion Engine: High-Speed Engines*.
11. Quoted in *Flying Magazine*, July 1935.

12. Setright, *The Power to Fly*.
13. In a letter to Lady Ottoline Morrell.

CHAPTER EIGHT

1. Rolls, *Steel Chariots in the Desert*.
2. Rolls thus contradicts those who suggest that Lawrence was a lone operator.
3. Lawrence, *Seven Pillars of Wisdom*.
4. Alan Bennett, *Forty Years On*, 1968.

CHAPTER NINE

1. F. Scott Fitzgerald, 'Echoes of the Jazz Age', *Scribner's Magazine*, 1931.
2. Wallace, *The Flight of Alcock and Brown*.

CHAPTER TEN

1. Harker, *The Engines Were Rolls-Royce*.
2. Ibid.
3. Saint-Exupéry, *Wind, Sand and Stars*, p. 000. Highly recommended.
4. From John Milton, *Areopagitica*, 1644.
5. Banks, *I Kept No Diary*.
6. Setright, *The Power to Fly*.
7. *The Times*, 27 November 1862.
8. B. J. Kraus, P. E. Godici, and W. H. King, *Reduction of Octane Requirement by Knock Sensor Spark Retard System*. SAE Technical Paper No. 780155 (1978).
9. Bill Bryson, *A Short History of Nearly Everything*, Black Swan, 2003.

10. Banks, *I Kept No Diary*.

11. F. Scott Fitzgerald, 'Echoes of the Jazz Age'.

CHAPTER ELEVEN

1. From Rudyard Kipling, 'The Female of the Species', 1911. One of Lady Lucy's favourite quotations.

2. Day, *Lady Houston*.

3. Boris Johnson, *The Churchill Factor*, Hodder & Stoughton, 2014.

4. MacNair, *Lady Lucy Houston*.

5. Day, *Lady Houston*.

6. Ibid.

7. Mitchell and Rodway, *Prelude to Everest: Alexander Kellas*, p. 000. A fascinating biography about an extraordinary man.

8. Sir Alan 'Tommy' Lascelles, *The King's Counsellor: Abdication and War – The Diaries of 'Tommy' Lascelles*, ed. Duff Hart-Davis, Weidenfeld & Nicolson, 2006.

9. Andrew Rose, *Scandal at the Savoy: The Infamous 1920s Murder Case*, Bloomsbury, 1991.

10. Rupert Godfrey, *Letters from a Prince: Edward to Mrs Freda Dudley Ward 1918–1921* (11 July 1920), Little, Brown & Co, 1998.

11. Philip Ziegler, *King Edward VIII*, Alfred A. Knopf, 1991.

12. Mitchell, *R. J. Mitchell – Schooldays to Spitfire*, History Press, 2006.

CHAPTER TWELVE

1. L. R. R. Fell, Rolls-Royce Owner's Club. https://rroc.org.au/wiki/index.php?title=Working_For_Mr_Royce.

2. Harker, *The Engines Were Rolls-Royce*.

3. Charles Jennings, *The Fast Set*, Abacus, Little, Brown Book Group, 2004.

4. A. J. P. Taylor, *English History 1914–1945*, Oxford, Clarendon Press, 1965.

5. Brendon, *Winston Churchill*.

6. Hansard, 21 October 1933.

7. Hansard, 7 February 1934.

8. Nockolds, *The Magic of a Name*.

9. Ovid, *Metamorphoses*, Book IV, line 428.

10. Rubbra, *Rolls-Royce Piston Engines*. The surname Rubbra is a corruption of the Old English 'ruh', meaning 'rough', or 'overgrown', and 'beorg', or 'hill'.

11. Rubbra, *Rolls-Royce Piston Engines*.

CHAPTER THIRTEEN

1. Harker, *Rolls-Royce from the Wings*.

2. Helmut Erfurt, *Junkers Ju 87 (Black Cross Volume 5)*, Bonn, Bernard & Graefe Verlag, 2004.

3. Ferran Bono, 'Las mortíferas pruebas de los "stukas" de Castellón', *El País*, 2 May 2018.

4. See John Weal, *Junkers Ju 87 Stukageschwader 1937–41*, Oxford, Osprey, 1997.

5. Jackson, *Infamous Aircraft*.

CHAPTER FOURTEEN

1. Pierre Clostermann.

2. Roald Dahl, *The Wonderful Story of Henry Sugar*, Cape, 1977.

3. According to Tony Butler, *British Secret Projects: Fighters and Bombers 1935–1950*, Midland Publishing, 2004.

4. Alfred, Price, *Spitfire: A Documentary History*.

5. Letter to Air Commodore Cave-Brown-Cave, August 1934.

6. Documentary *Spitfire*, copyright Elliptical Wing Ltd, 2019. Also see the *Journal of Aeronautical History Paper No. 2013/02 121 The Spitfire Wing Planform: A Suggestion*, by J. A. D. Ackroyd. Fascinating stuff.

7. F. W. Meredith, *Cooling of Aircraft Engines. With Special Reference to Ethylene Glycol Radiators Enclosed in Ducts*, Aeronautical Research Council, 1936.

8. Morgan and Shacklady, *Spitfire: The History*, 5th rev. edn, London, Key Publishing, 2000.

9. Webb, *Never a Dull Moment*.

10. Clostermann, *Flames in the Sky*.

11. Carles Boix, *Democratic Capitalism at the Crossroads*, Princeton University Press, 2019.

12. http://www.nationalarchives.gov.uk/first-world-war/home-front-stories/strike-action/

13. https://aviation-safety.net/wikibase/wiki.php?id=161510

CHAPTER FIFTEEN

1. Strachey, *Eminent Victorians*.

2. Letter to Hilda Chamberlain, 30 December 1939.

3. Self, *Neville Chamberlain*.

4. Webb, *Never a Dull Moment*, Vickers Archives 508, Cambridge University Library.

5. Clostermann, *The Big Show*.

6. The ability of this remarkable engineer was recognised by his title: 'Dr h.c.mult.' – *Doctor honoris causa multiplex* – holder of many honorary doctorates.

7. George Orwell, 'Politics and the English Language', *Horizon*, April 1946.

CHAPTER SIXTEEN

1. Carl von Clausewitz, *Vom Kriege*.
2. The Party member was Friedrich Wiedemann, quoted in Thomas Weber, *Hitler's First War*, OUP Oxford, 2011.
3. Sir Neville Henderson, *Failure of a Mission*, G. P. Putnam's Sons, 1940.
4. See Taylor, *Boxkite to Jet*, the story of Frank Halford's engineering from First World War biplanes to the Ghost jet. Recommended.
5. Clostermann, *The Big Show*.

CHAPTER SEVENTEEN

1. Saint-Exupéry, *Wind, Sand and Stars*.
2. http://www.wwiiaircraftperformance.org/mosquito/w4076. pdf
3. Horst Boog, Gerhard Krebs and Detlef Vogel, *Germany and the Second World War: Volume VII: The Strategic Air War in Europe and the War in the West and East Asia, 1943–1944/5*, Oxford, Clarendon Press, 2006.
4. Pilot Officer Max Sparks.
5. Wooldridge, *Low Attack*.

CHAPTER EIGHTEEN

1. Hosea 8:7.
2. See Kirby, *The Avro Manchester*: the Vulture's shortcomings explained in technical detail.
3. Lettice Curtis, quoted in Iveson, *The Lancaster and the Tirpitz*.

4. Noël Coward, 'Lie in the dark and listen', *Collected Poems*, Methuen, 1984.

5. https://www.cia.gov/library/center-for-the-study-of-intelligence/csi-publications/csi-studies/studies/95unclass/Leary.html

6. Harker, *The Engines Were Rolls-Royce*.

7. http://aircrewremembered.com/Nighthawks/nighthawks50.html

8. See Ezra Bowen, *Knights of the Air (Epic of Flight)*, New York, Time-Life Books, 1980.

CHAPTER NINETEEN

1. Harker, *The Engines Were Rolls-Royce*.

2. Setright, *The Power to Fly*.

IN MEMORIAM

1. Alfred, Lord Tennyson, *In Memoriam*, lines 27–8.

BIBLIOGRAPHY

Banks, Rod, *I Kept No Diary*, Crawley, Airlife Publishing Ltd, 1978

Bentley, W. O., *My Life and My Cars*, London, Hutchinson & Co, 1967

Birch, David, *Rolls-Royce and the Mustang*, Derby, Rolls-Royce Heritage Trust, 1987

Brendon, Piers, *Winston Churchill: A Brief Life*, Oxford, Isis, 1986

Clarke, Tom C., *Royce and the Vibration Damper*, Derby, Rolls-Royce Heritage Trust, 2003

Clostermann, Pierre, *Flames in the Sky*, London, Chatto & Windus, 1952

—, *The Big Show: The Greatest Pilot's Story of World War II*, London, Weidenfeld & Nicolson, 2004

Craighead, Ian, *Rolls-Royce Merlin 1933–50: Owner's Workshop Manual*, Sparkford, Haynes, 2015

Day, J. Wentworth, *Lady Houston: The Richest Woman in England*, London, Allan Wingate, 1958

Dee, Richard, *The Man Who Discovered Flight: George Cayley and the First Airplane*, Toronto, McClelland & Stewart Ltd, 2007

Deighton, Len, *Bomber*, London, Cape, 1970

Grahame, Kenneth, *The Wind in the Willows*, London, Methuen, 1908

Harker, Ronald W., *Rolls-Royce from the Wings: Military Aviation, 1925–71*, Oxford, Oxford Illustrated Press, 1976

—, *The Engines Were Rolls-Royce*, New York and London, Macmillan, 1979

Heilig, John, *Rolls-Royce: The Best Car in the World*, London, Apple Press, 2000

Hodges, Andrew, *Alan Turing: The Enigma*, New York, Simon & Schuster, 1983

Howard, Fred, *Wilbur and Orville: A Biography of the Wright Brothers*, New York, Dover, 1987

Hoyland, Graham, *Last Hours on Everest*, London, HarperCollins, 2014

Iveson, Tony, *The Lancaster and the Tirpitz: The Story of the Legendary Bomber and How it Sunk Germany's Biggest Battleship*, London, André Deutsch, 2014

Jackson, Robert, *Infamous Aircraft – Dangerous Designs and Their Vices*, Barnsley, Pen & Sword, 2005

Kirby, Robert, *The Avro Manchester: The Legend Behind the Lancaster*, Stroud, Fonthill, 2015

Lawrence, T. E., *Seven Pillars of Wisdom* (1926 Subscribers' Edition)

Lewis, Cecil, *Sagittarius Rising*, London, Peter Davies, 1936

Lloyd, Ian, *Rolls-Royce: The Growth of a Firm*, London, Macmillan, 1978

—, *Rolls-Royce: The Merlin at War*, London, Palgrave Macmillan, 1978

Lyman, Robert, *The Jail Busters: The Secret Story of MI6, the French Resistance and Operation Jericho*, London, Quercus, 2014

McKinstry, Leo, *Spitfire, Portrait of a Legend*, London, John Murray, 2007

McManus, Peter, *Motorcycles, Merlins and Mosquitos: The Story of Chris Harrison, Racing Motorcyclist, Rolls-Royce Engineer, Mosquito Pilot*, Derby, Breedon, 2009

MacNair, Miles, *Lady Lucy Houston, DBE*, Barnsley, Pen & Sword, 2016

Mitchell, Dr Gordon, *R. J. Mitchell – Schooldays to Spitfire*, Cheltenham, History Press, 2006

Mitchell, Ian R., and George, W. Rodway, *Prelude to Everest: Alexander Kellas, Himalayan Mountaineer*, Edinburgh, Luath Press, 2011

Montagu of Beaulieu, Lord, *Rolls of Rolls-Royce*, London, Cassell, 1966

Morgan, Eric B., and Edward Shacklady, *Spitfire: The History*, 5th rev. edn, Key Publishing, 2000

Needham, Joseph, *Science and Civilisation in China*, Cambridge, Cambridge University Press, 1956

Nockolds, Harold, *The Magic of a Name*, London, G. T.Foulis, 1959

Pugh, Peter, *The Magic of a Name*, Duxford, Icon, 2000

Ricardo, Sir Harry R., *The Internal Combustion Engine: High-Speed Engines*, London, Blackie, 1923

Rolls, Sam, *Steel Chariots in the Desert*, London, Jonathan Cape, 1937

Rubbra, A. A., *Rolls-Royce Piston Engines*, Derby, Rolls-Royce Heritage Trust, 1990

Saint-Exupéry, Antoine de, *Wind, Sand and Stars*, New York, Reynal & Hitchcock, 1939

Self, Robert, *Neville Chamberlain: A Biography*, Aldershot, Ashgate, 2006

Setright, L. J. K., *The Power to Fly*, London, Allen & Unwin, 1971

Strachey, Lytton, *Eminent Victorians*, London, Chatto & Windus, 1918

Taulbut, Derek S., *Eagle: Henry Royce's first aero-engine*, Derby, Rolls-Royce Heritage Trust, 2011

Taylor, Douglas, *Boxkite to Jet*, Derby, Rolls-Royce Heritage Trust, 1999

Wallace, Graham, *The Flight of Alcock and Brown*, London, Putnam, 1955

Webb, Denis le P., *Never a Dull Moment at Supermarine*, Hellingly, J & KH Publishing, 2001

Wilson, Gordon A. A., *The Merlin: The Engine that Won the Second World War*, Stroud, Amberley, 2018

Wooldridge, John, *Low Attack: The Story of Two Mosquito Squadrons, 1940–43*, London, Crecy, 1993

IMAGE CREDITS

INDEX

40/50 Silver Ghost 47, 66–67, 113, 123, 149, 150
617 Squadron, RAF 306–9

'Aerodrome, The' 14–17, 18
Airco DH2 aircraft 103
air-cooled engines, benefits of 40–42, 92–93
airships 106, 108
Albatros D.1 aircraft 109
Alcock, Jack 126, 128–29
Alkemade, Flight Sergeant Nicholas 312–13
Alliance P.2 Seabird aircraft 130
Allison V12 engine 315
Antoinette 8V engine 38–40
Archdeacon, Ernest 32, 33
Archer Jr, Lee 316
Arkwright, William 78–79
Armstrong Siddeley engines 139
Arnold, Thomas 48
Austin Seven 78
Austrian Alpine Trial 84–85
Aviatek spotter aircraft 100–101
Avro 504 aircraft 100
Avro Lancaster aircraft 3, 208–9, 301–313
Avro Shackleton aircraft 322
Avro Stratosphere aircraft 320–21
Avro Tudor aircraft 331

Baldwin, Stanley 204, 249
Balzer, Stephen 15
Banks, Rodney 151, 153–54, 168
Baracca, Francesco 105
Barkers carriage builders 71, 79

Barnato, Diana 234–35
Battle of Britain (1940) 3, 229, 252, 261, 263–64, 271–87
Battle of Graveney Marsh (1940) 278–79
Beaverbrook, Lord 249–50, 269, 324
Bentley engines 92–93, 142
Bentley, Walter Owen 92–93, 97, 183
Bevin, Ernest 232, 233
Biard, Henri 142
Billing, Noel Pemberton 134
Blackburn Kangaroo aircraft 130
Blacker, Lieutenant-Colonel Latham Valentine 171
Blanchard, Jean-Pierre 7
Blériot, Louis 125
Bluebird 179
BMW engines 182, 260, 319
Boeing B-17 Flying Fortress aircraft 290
Bollee motorised tricycle 61
Boorman, Warrant Officer Bob 317
Boothman, Flight Lieutenant John 168
Boulton Paul Atlantic aircraft 126
Brinckman, Lieutenant-Colonel Sir Theodore 159
Bristol Aeroplane Company 236, 322–23
Bristol engines 133, 139, 172, 236, 323–25
Bristol fighter aircraft 105
Brown, Arthur 126–29
Brunel, Sir Marc 267
Bryant, Sergeant Bill 317
Bryson, Bill 152

Buchan, Colonel John 171
Bulman, Major George 143, 144, 193, 213
Buzzard engines 145
Byron, George Frederick William 159–60

Camm, Sydney 211–13
Canadair North Star aircraft 331
Cantopher, Captain John 279
Carrick, Sergeant Mike 292–93
Cassell, James 148–49
Castoldi, Mario 166
Cayley, Sir George 8–10, 22
Cecil, Robert 125
Chadwick, Roy 305, 320–21
Chamberlain, Neville 173, 193, 232, 247–48
Christmas Bullet Flexible Aeroplane 110–11
Christmas, Dr William 110–11
Churchill, Winston 115, 118, 120, 129, 184–87, 232, 233, 249, 252, 262–64, 271, 286–87, 301–302
Citroën, André 191–92
Claremont, Ernest 56, 90
Clausewitz, Carl von 271
Clostermann, Pierre 226, 257–58, 283–85, 325–26
Clydesdale, Lord 171
Condor engines 144–45
Convair B-36 Peacemaker aircraft 327
Corncob engine 327
Cosmos Jupiter engine 323
Coverley, Bob 180
Coward, Noël 249, 273
Crusader tanks 317
Curtis, Lettice 305
Curtiss D-12 engine 137, 138–39, 142
Curtiss, Glenn 18
Curzon, Lord 162

Dahl, Roald 214
Daimler engines 85–86
Daimler, Gottlieb 13
Daimler Mercedes engines 95, 96
Daimler-Benz engines 189, 254–57, 258–61, 279–80, 283, 287
Dambusters, The 306–7

Daniels, John T. 26, 30
De Dion motorised tricycle 61
De Dion-Bouton 53
de Havilland Comet aircraft 332
de Havilland, Geoffrey 289–90, 292, 293
de Havilland Mosquito aircraft 3, 289–99
Decauville 54, 55–56
Deperdussin monoplanes 136
Dornier 217 aircraft 320
Dornier Do 17 aircraft 183, 277
Dowding, Sir Hugh 218, 219, 271
Dragonfly engines 107

Eagle engines 90–98, 104–5, 106, 108–9, 126–29, 326
Edison, Thomas 55
Edmunds, Henry 49, 67
Eilmer of Malmesbury 7
Einstein, Albert 5, 21
electric cars 54
Elliott, Albert 194–95, 303
English Electric Lightning T4 aircraft 235
Enigma Codes 269
Enlightenment, The 43–44
Eyston, Captain George 179–80

F.2 fighter aircraft 104, 109
Fairey Battle aircraft 197, 198, 216, 252, 253
Fairey Fox aircraft 138, 140–41
Fairey, Richard 137–41
Fairey Special aircraft 126
Falcon engine 98, 105
Farman III biplanes 125
FE2d fighter aircraft 98, 99
Fedden, Sir Roy 106, 236
Felixstowe Flying Boats 109
Ferber, Captain Ferdinand 32
Ferrari engines 166–67, 168
Ferrari, Enzo 105
Fiat engines 166, 168
First World War (1914–18) 89–90, 93, 97–98, 99–111, 150–51
  Battle of the Somme (1916) 103–4
  Lawrence of Arabia 114–21
  Sykes-Picot Agreement (1916) 114
Fitzgerald, F. Scott 123, 155

Foch, Marshal Ferdinand 181
Focke-Wulf FW 190 aircraft 256, 260, 298, 319–20
Fokker Eindecker E.III aircraft 100, 102–3
Ford, Henry 264
fuel quality, importance of 147–53, 260–61, 280
Fulton Jr, Robert Edison 314

Galileo 43–44
Galland, Adolf 182
Ge Hong 6
General Motors engines 149
Giffard, Henri 7
Gilbert, Squadron Leader 234–35
Gill, Eric 177–78
Giotto 221
Gladiator aircraft 274
Glinert, Ed 49–50
Gnome rotary engines 40–42, 103, 136, 142
Gordon Bennett Race 62
Göring, Hermann 182–83, 281, 285, 295–96, 317
Goshawk engine 98, 219
Grahame, Kenneth 64–65
Graveney Marsh, Battle of (1940) 278–79
Graves, Robert 119
Green, Gustavus 43
Gretton, Frederick 159
Grey, Sir Edward 89
Griffon engine 188, 321–22

Handley Page 0/100 aircraft 98, 99, 104, 109
Handley Page 0/400 aircraft 108
Handley Page Halifax aircraft 314
Handley Page V/1500 aircraft 108, 109, 126, 128
Harker, Ronald 139–40, 204–5, 315–16, 319
Harmsworth, Alfred 124–25
Harris, Sir Arthur 'Bomber' 306, 313
Hawk engine 98, 106
Hawker Fury aircraft 180
Hawker, Harry 126–27, 211
Hawker Hart aircraft 140, 141, 180, 194, 204–5, 212

Hawker Hurricane aircraft 2, 197, 212–15, 245, 253, 274, 277
Hawker Tempest aircraft 326
Hawker Typhoon aircraft 325
Heinkel He 51 aircraft 182
Heinkel He 70 Blitz aircraft 182
Heinkel He 111 aircraft 183, 251, 259, 277
Henshaw, Alex 208–9
Herbert, Audrey 121
Hillary, Richard 226
Hitler, Adolf 54, 161, 163, 173, 182, 193, 198, 254, 265, 272, 286, 287, 293
Hives, Ernest 67–68, 85, 93, 154, 228–29, 230, 264, 269, 316
Hooke, Robert 8
Hooker, Stanley 230
horsepower, relevance of 12–13, 78, 251, 282, 325
Houston, Lady Lucy 3, 157–61, 163–73
Houston, Sir Robert 160–61, 163–65
HZ-Anlagen system 320

Industrial Revolution 44
internal combustion piston engine, invention of 6, 9, 10–18
inter-war years (1919–38) 123–31, 143–55, 157–74, 177–98, 201–9, 211–28
Italo-Turkish War (1911) 108

jet engines 327–30
Johns, W. E. 118
Johnson, Boris 161
Johnson, Claude 67, 68, 79, 81, 84, 85, 91, 95–96, 100, 106, 124, 126
Jolly, Lieutenant Allington Joyce 111
Jumo J211 engine 259
Junkers Ju 87 Stuka aircraft 183, 205–6, 252, 277
Junkers Ju 88 aircraft 251, 278
Jupiter engine 133

Kellas, Alexander 170–71
Kestrel engine 98, 139–41, 147, 179–80, 205–6, 279
Keynes, John Maynard 181
Kindelberger, James Howard 'Dutch' 314–5

Kipling, Rudyard 14
Kitty Hawk Flyer 18, 19, 21–26

Langley, Professor Samuel 13–18, 22, 26
Lawrence, D. H. 108–9
Lawrence of Arabia 114–21, 135, 154
Lenin, Vladimir 80–81
Lenoir, Jean Joseph Étienne 10
Leopold II of Belgium 80
Levavasseur, Léon 38–39
Lewis, Cecil 103–4
Liberty engines 110, 111, 150, 265
Lilienthal, Otto 22
Lindbergh, Charles 129
Lindemann, Professor Frederick 185–86
List, Hans 261
London Science Museum 18–19, 31
Lord, Frank 56–57
Lovesey, Cyril 153, 237
Low, David 273
Lucas, Philip 213

Macchi M.39 aircraft 142–43
Macchi MC72 aircraft 166, 168
MacDonald, Ramsay 165, 169, 185
MacGill, Elsie 215
Manly, Charles 14–18
Martinsyde biplanes 101–2, 126
Martinsyde Raymor aircraft 126
Martinsyde Type A aircraft 130
Masaryk, Jan 248
Maybach, William 13
McLean, Sir Robert 223
Mercedes engines 43, 189
Meredith, Frederick 222
Merlin engines 98, 157, 188–98, 204–5, 212–15, 219, 221–24, 227–32, 240–47, 251–61, 264–65, 268, 271–83, 287, 289, 290, 292, 293, 301, 302, 303, 304–17, 319, 320, 331
Messerschmitt Bf 109 aircraft 183, 245, 251, 252, 254, 256, 260, 261, 277, 279–85, 317
Messerschmitt Bf 110 aircraft 281
Messerschmitt Me 210 aircraft 207–8
Messerschmitt Me 262 aircraft 326, 329

Meteor engine 315
Midgley Jr, Thomas 151–53
Miller, Bill 232
Mills, Cuthbert 111
Mitchell, Dr Gordon 174
Mitchell, Reginald 133–35, 141–44, 154, 216–21, 223, 225, 226–27
Montagu, Lord 61, 62, 63, 64, 66, 73–74, 82
Morane scout aircraft 103–4
Mosley, Sir Oswald 162
Motor Volunteer Corps 72
Mount Everest high-altitude flight 170–72
Mussolini, Benito 142, 161, 163, 165–66, 173, 198

Napier cars 68–69
Napier engines 126, 133, 137, 138, 142, 183, 323–27
Napier, Sir Charles 267
Niépce, Claude 9
Niépce, Nicéphore 9
Nieuport 11 aircraft 103
Nockolds, Harold 2
Nomad engine 326–27
Nordenfeldt engines 95
Northcliffe, Lord 124–25

Ohain, Hans von 329, 330
Operation Jericho (1944) 296–98
Orlebat, Squadron Leader Augustus 180
Orwell, George 264
Otto, Nikolaus 10, 11, 13
Overstreet, Bill 317

P-51 Mustang aircraft 314–17
Packard engines 265
Packard Motor Company 265–66, 268–69
Panhard, René 55
Paris-Toulouse-Paris race 69
Pasnofsky, Professor Erwin 81–82
Paulhan, Louis 125
PB-1 Supermarine flying boats 134
Pegasus engine 172
Pénaud, Alphonse 8, 21
Peterloo Massacre (1819) 231
Peugeot Phaeton 61, 63–64

Picasso, Pablo 202–3
Pickard, Group Captain Percy Charles 296–98
Pomeroy, Lawrence 84
Porsche, Ferdinand 54
Pratt & Whitney engines 325
propellers 25, 235–37

Quatorze-bis (14b) 39–40
Quill, Jeffrey 221, 224–25, 226

R engine 144–55, 165, 167–69, 177, 179, 180
Radley, James 84–85
Renault engines 43, 92, 94
Ricardo, Harry 323
Ricardo, Henry 104–5
Richthofen, Wolfram Freiherr von 201–2
Riley Nine 78
Rolls, Charles 47, 49–50, 58–74, 174, 183
Rolls, Sam 113, 114, 115–16
Rolls-Royce
40/50 Silver Ghost 47, 66–67, 77–85, 113, 123, 149, 150
Buzzard engine 145
Condor engine 144–45
Eagle engine 90–98, 104–5, 106, 108–9, 126–29, 326
Falcon engine 98, 105
during the First World War (1914–18) 89–90, 99–109, 113, 116–18, 150–51
Goshawk engine 98, 219
Griffon engine 188, 321–22
Hawk engine 98, 106
the inter-war years (1919–38) 123, 126–29, 130–31, 133–37, 139–41, 143–55, 177–81, 187–98, 204–6, 212–15, 219, 221–25
Kestrel engine 98, 139–41, 147, 179–80, 205–6, 279
Merlin engine 157, 188–98, 204–5, 212–15, 219, 221–24, 227–32, 251–61, 264–65, 268, 271–83, 287, 289, 290, 292, 293, 294, 301, 302, 304–17, 319, 320, 331
Meteor engine 317
origins of the company 49–74

Phantom V12 77
R engine 144–55, 165, 167–69, 177, 179, 180
Schneider Trophy races 133–37, 141–44, 151, 153–55, 167–69
during the Second World War (1939–45) 227–32, 233–34, 237, 251–61, 264, 265, 267, 271–76, 319–22
Silver Shadow 79–80
take-over of Bentley 183–84
Vulture engine 302–4, 314
Rolls-Royce Armoured Car 72
Rolls-Royce Phantom V12 77
Roosevelt, Franklin D. 119, 233
Rotol Airscrews Ltd 236–37
Rowledge, Arthur 126, 140, 188, 191
Royal Aircraft Factory 94
Royce, Henry 14, 47, 49–59, 68–73, 91–92, 94–97, 144, 146, 154, 155, 177–78, 180–81, 187, 190–91, 197
Rubbra, Arthur 192–93, 195
Ryneveld, Lieutenant Colonel van 131

SA8/75 engine 42–43
Sabre engine 321–24
Saint-Exupéry, Antoine de 141
Schneider, Jacques 135–36
Schneider Trophy races 120, 133–37, 141–44, 151, 153–55, 158, 165–68, 174
Schwager, Hans 100
Scott, Lord Herbert 91
Scottish Reliability Trials 82, 84
Scott-Paine, Hubert 135
Sea Lion II biplanes 137
Seagrave, Sir Henry 179
Second World War (1939–45) 2–3, 228–37, 239, 263–67, 289–99, 301–317, 319–30
617 Squadron, RAF 306–8
Battle of Britain (1940) 3, 229, 252, 259, 261, 263–64, 271–87
Battle of Graveney Marsh (1940) 278–79
Dambusters, The 306–7
Operation Jericho (1944) 296–98
Seguin, Louis and Laurent 40
Selfridge, Lieutenant Thomas 32
Setright, L. J. K. 17, 107, 143–44, 183, 324, 325–26

Shenstone, Beverley 217, 220, 221, 225
Sherira, Ben 6
Shilling, Beatrice 245–46
Short Brothers 95
Short Shamrock aircraft 126
Sidgreaves, Arthur 135
Silver Ghost 47, 66–67, 113, 123, 149, 150
Silver Shadow 79–80
Six, SS-Brigadeführer Dr Franz 273, 274
Small, Lieutenant Freddy 100–101
Smithsonian, the 13–14, 18–19, 22, 26
Snow, C. P. 48
Somme, Battle of the (1916) 103–4
Sopwith Atlantic aircraft 126–27
Sopwith Camel aircraft 93
Sopwith Schneider biplanes 136
Sopwith Wallaby aircraft 130
Spanish Civil War (1936–39) 183, 198, 201–3
Speed of the Wind 179–80
Speed Spitfire 227–28
speedboats 38–39, 179
Speer, Albert 272, 307
Spirit of Ecstasy, The 82
Spitfires *see* Supermarine Spitfire aircraft
Stainforth, Flight Lieutenant George 169
Stalin, Joseph 233
steam engines 5–6
Strachey, Lytton 239, 273
Strange, Flight Commander Louis 100–102, 253–54
Summers, Joseph 'Mutt' 224
Supermarine Aviation Works 133–34, 141–43, 165
Supermarine S4 aircraft 142
Supermarine S5 aircraft 133, 143
Supermarine S6 aircraft 133, 146, 154–55, 167–69, 174, 180, 217
Supermarine Spitfire aircraft 2, 157, 216–28, 237, 245, 249–50, 251, 274–76, 278, 281, 283–85, 292, 316, 319, 320, 322, 331
Sykes, Charles Robinson 82
Sykes–Picot Agreement (1916) 114

Tank, Kurt 319
tanks 317
Tatra T97 54–55
Taulbut, Derek 109
Taylor, A. J. P. 184
Taylor, Charlie 23–24, 25, 26
Thomas, Lowell 118
Thousand Mile Reliability Trial 62
Thunderbolt 179, 180
Tirpitz battleship 301, 307–9
transatlantic flights 126–29
Treaty of Versailles (1919) 181, 254
Trenchard, Air Chief Marshal Hugh 138
Trouvé, Gustave 8
Tryon, H. C. 323
Turing, Alan 48

Udet, Ernst 276

valveless engines 85–86
Vauxhall 62
Vickers Vimy 128–29
Vickers Vimy aircraft 126, 130–31
Vinci, Leonardo da 7, 44
Volkswagen Beetles 78
Vulture engine 302–304, 314

Wallis, Barnes 306, 307, 309
water-cooled engines 40, 94
Watt, James 12
Weir, Sir William 107
Wells, H. G. 263
Wendel, Fritz 207–8
Wenxuan, Emperor 6–7
West, Rebecca 273
Whittle, Frank 327–29
Whitworth, Sir Joseph 266
Willis, Corporal George 279
Wooldridge, Wing Commander 298–99
Wooler, Ernest 57
Wright Flyer 21–26, 29–34, 72–73, 331
Wright, Orville 17, 18–19, 21–26, 29–34, 37–38, 72
Wright, Wilbur 17, 21–26, 29–34, 37–38, 72

Young, James 'Paraffin' 148